To David
From All ST
OCT.

Ellis Rowan

A Flower-Hunter in Queensland

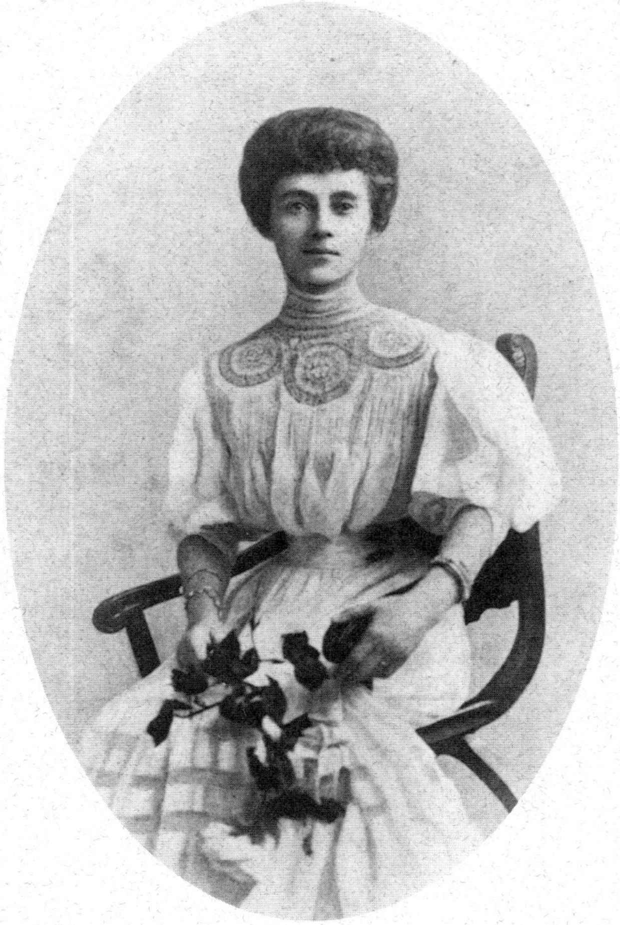

Ellis Rowan c. 1910, just before she began her second series of visits to Queensland. The English flower painter Marianne North called Ellis "a very pretty fairy-like little woman, always over-dressed". (*Lone Hand*, October 1910)

Ellis Rowan
A Flower-Hunter in Queensland

Judith McKay
Queensland Museum

PROUDLY PRINTED IN QUEENSLAND

Requests for this book should be made to: Queensland Museum
PO Box 300 South Brisbane Queensland 4101 Australia.

National Library of Australia
Cataloguing-in-Publication data.

McKay, Judith Marilyn, 1949-
Ellis Rowan, A Flower-Hunter in Queensland.

Bibliography.
ISBN 0 7242 3847 6.

1. Rowan, Ellis, 1848-1922 — Exhibitions. 2. Rowan, Ellis, 1848-1922 — Biography. 3. Queensland Museum — Exhibitions.
4. Flower painting and illustration — Queensland — Exhibitions.
5. Wild flowers in art — Exhibitions.
6. Painting — Queensland — Brisbane — Exhibitions
7. Painters — Australia — Biography.
I. Rowan, Ellis, 1848-1922. II. Queensland Museum. III. Title.

759.9940749431

First printed 1990

Designed and produced by: Margy Lukeij

Editor: Michelle Ryan

Typeset in: Century Expanded 10/13pt

Printed on Speckletone Chalk White 104gsm and Ikonorex Dull 150gsm by: Fergies Colour Printers, Brisbane.

This book was compiled on a Macintosh SE Desktop publishing system kindly donated in part by Apple Computer Australia Pty Ltd.

Cover: Detail of cat. no. 36 from the Queensland Museum's collection of 125 paintings by Ellis Rowan.

CONTENTS

List of Illustrations

PREFACE

The Queensland Museum's collection of 125 flower paintings by Ellis Rowan has been inaccessible to the public for almost half a century. This book marks the conservation and exhibition of the collection, made possible by a generous grant from the Queensland Government. The conservation work, apart from preserving the collection for posterity, has enabled the paintings to be catalogued and studied with the aid of evidence now revealed on the reverse of many of the paintings.

It is a rare privilege to curate an outstanding collection of paintings and to find behind them a fascinating personality. As a contemporary of Ellis Rowan's concluded:

> To those fortunate enough to meet her, the painter was more wonderful than her work — and that is saying a great deal.

Ellis Rowan was an enigmatic woman far ahead of her time. She maintained a successful and independent career as an artist while indulging her love of flowers and picturesque travel. A shameless self-publicist, she wrote of her life's adventures and enjoyed holding her listeners spellbound with stories of how she had dangled from precipices and travelled among head-hunters to find subjects for her paintings. But to sort the fact from the fiction and to piece together a chronology of her life is a much harder task, admirably attempted by Margaret Hazzard in her biography *Australia's Brilliant Daughter Ellis Rowan: Artist, Naturalist, Explorer 1848-1922*, published in 1984.

By searching archival sources and old newspapers I have been able to reconstruct six visits by the artist to Queensland — instead of the two previously known — and to discover the circumstances of the acquisition of her Queensland collection. Where possible, I have added dates and identities to Ellis's own graphic account, *A Flower-Hunter in Queensland and New Zealand*, published in 1898. Her account is, in fact, a composite of her three 19th century visits to Queensland, simplified to appear as only one extensive journey and mentioning her hosts and hostesses mostly only by an initial ("Mrs A.", etc.). Ellis's cuttings book, held by the National Library of Australia, also presented a challenge for unfortunately, newspaper dates and titles were not clipped along with the articles. Likewise, her paintings are mostly undated. Ellis has left no descendants, nor diaries, nor a comprehensive collection of letters. Consequently, some of her reported travels in Queensland remain a mystery. I have not been able to date her visits to Charters Towers and Hughenden, and suspect her visits to Georgetown, Croydon and across the Northern Territory to be — in true Ellis fashion — more fictional than factual.

The Ellis Rowan project could not have been undertaken without the co-operation and assistance of many institutions and individuals. Thanks are due to the following:

The Queensland Herbarium, Department of Primary Industries, especially Mr Rod Henderson; the State Library of Queensland, especially the John Oxley Library and the microfilm reading room; Queensland State Archives; Queensland Art Gallery especially the conservation staff; Queensland Railways Historical Section; Fryer Memorial Library, University of Queensland; the Royal Historical Society of Queensland, especially Mrs Lyn Love; the Queensland Women's Historical Association, especially Miss Gyneth Campbell; Brisbane City Hall Art Gallery and Museum; the National Library of Australia, Canberra, especially the Pictorial Section; the Mitchell Library, Sydney; the Art, Music and

Performing Arts Library of the State Library of Victoria; the Powerhouse Museum, Sydney; the Archives of Business and Labour, Australian National University, Canberra; the Library of the Royal Botanic Gardens, Kew; and Cambridge University Library.

Queensland Museum volunteers gave generous assistance with the historical research: Miss Diana Post, Miss Donna Miles, Mrs Betty Downing and Mrs Beryl Wilson. Volunteers Mr John and Mrs Margaret Parham provided valuable botanical information and advice. The following individuals provided other useful information, mostly in response to public appeals kindly published by newspapers throughout Queensland: Mrs Midge Allom, Miss Helen Annand, Mr Philip Bacon, Mr Archie Black, Mrs Kay Brown, Lady Campbell, Mr Charles A. Campbell, Dr and Mrs Drury Clarke, Mrs Lesley Cobon, Ms Kate Collins, Mr Glenn Cooke, Mr Peter Croker, Ms Fiona Gardiner, Dr A. J. Harrison, Mrs Carol Huddy, Mr Duncan Jackson, Capt. Graham Jardine-Vidgen, Mrs J. Jefferies, Mrs Joan Karlino, Mrs Nancy Keable, Dr Joan Kerr, Mr John Kerr, Dr Ruth Kerr, Mrs Doris Lakeland, Mr Tim Low, Mrs Margo Macarthur, Mrs Jane MacGowan, Mrs W. D. McKenzie, Dr Cathryn Mittelheuser, Mrs Ann Neale, Mrs Majorie Schumacher, Miss Enid Tardent, Mrs Helen Vellacott, Mr Don Vernon, Mr Blair Wilson, and the family of Ellis Rowan. Many of my Museum colleagues provided expertise and support, notably: Ms Michelle Ryan, Dr Geoff and Mrs Sybil Monteith, Mr Ian Sanker, Miss Danielle McDonald, Miss Margy Lukeij, Mrs Bronwyn Searle, Mrs Diana Flaherty, Dr Carden Wallace; Mr Stephen Van Dyck, Mr Paul Ramsden, Mr Bruce Cowell and the Museum's photographic staff. The painting collection was expertly conserved by Mrs Kerry McInnis and associates of Art and Archival Pty Ltd, Queanbeyan, and by the Queensland Art Gallery.

The following abbreviations are used in the book:

Flower-Hunter	Ellis Rowan, *A Flower-Hunter in Queensland and New Zealand*, 1898;
QSA	Queensland State Archives;
NLA	National Library of Australia.

Paintings in the Queensland Museum collection are referred to by their catalogue numbers (cat. no. –).

J.M.

Postscript: The Queensland Museum would be pleased to hear from any readers who can add new information to this story.

B I O G R A P H Y

Marian Ellis Rowan, always known as Ellis, was internationally acclaimed as a flower painter. She turned what her fellow Australian artists deemed an inferior art form — a "genteel female pastime"[1] — into an adventurous and profitable career which took her all over the world and even to the then remote tropics of Queensland.

Ellis, born in Melbourne on 30 July 1848, was the eldest child of pastoralist and later stock and station agent, Charles Ryan and his wife Marian, née Cotton. Her grandfather John Cotton had written and illustrated a book on English birds before settling in the Port Phillip District in 1843 as a pioneer pastoralist. The noted Le Souef family of zoologists were Ellis's cousins of the Cotton line. As the daughter of "the wealthy Ryan of Victoria"[2] Ellis inherited a privileged position in colonial society, as well as a talent for art and natural history.

Following her education in Melbourne at a school for young ladies, Ellis visited relatives in England in 1869-70. She probably received private art lessons there, though she always claimed to be entirely self-taught. Her paintings of native flora won a medal at Melbourne's International Exhibition of 1872, the first of the ten gold, fifteen silver and four bronze medals she received at international exhibitions of art and industry during the next twenty years. Ellis won a gold medal at Melbourne's International Exhibition of 1880 and was the only Australian artist to do so at the city's Centennial International Exhibition eight years later. Her successes aroused the envy of Melbourne's male professionals. After her 1888 win, the Victorian Artists' Society expressed "strong dissatisfaction" that the work of an amateur flower painter should have been rated above their more worthy figure studies and landscapes. Moreover they considered her gold medal "a direct insult" to the foreign artists who had competed.[3] From then on Ellis operated outside the world of artistic professionalism, and remained conscious of her limitations beyond her chosen domain of flower painting.

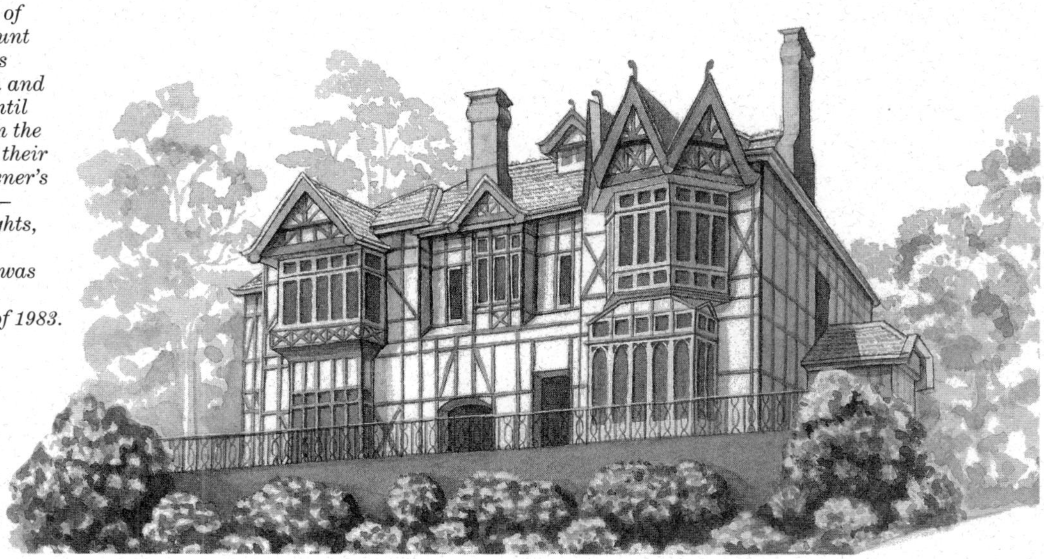

An artist's impression of Derriweit Heights, Mount Macedon, built by Ellis Rowan's father in 1874 and owned by the Ryans until 1896 when a reversal in the family fortunes forced their move to the head-gardener's house — The Cottage — nearby. Derriweit Heights, one of Victoria's finest boom-time residences, was destroyed in the Ash Wednesday bushfires of 1983.

In 1873 Ellis married Frederic Charles Rowan (1844-1892), a British army officer then serving in the New Zealand Armed Constabulary and a hero of the Maori Wars. She returned with him to his garrison at Pukearuhe near New Plymouth where, cut off from "a life full of gaiety" in Melbourne, she perfected her flower painting technique and set out on her ultimate goal of "getting as good a collection of the Australasian flora as I could".[4] Her husband, a keen amateur botanist and disciplinarian, spurred on her burgeoning career with advice and stern criticism.[5] The Rowans' only child, Eric, was born in 1875.

In 1877 they returned to Melbourne where Frederic Rowan took up a successful business career, becoming general manager of the Australasian Electric Light Power and Storage Company and Australian representative of the Pacific Telegraph Company. He was also a director of the company which in 1886 published the *Picturesque Atlas of Australasia*, with illustrations by Ellis and many other Australian artists. The company later intended to publish a handsome serial edition on Australasian flora.[6] This was to combine chromolithographs of Ellis's paintings with botanical text by Baron Ferdinand von Mueller (1825-1896), the esteemed Government Botanist of Victoria,[7] but the proposal resulted only in coloured supplements in the *Australian Town and Country Journal* Christmas editions of 1888 and 1889. Von Mueller was a family friend and ally, having been the first adviser to Ellis's father in the planning of the famous garden which surrounded the Ryan mansion at Mount Macedon northwest of Melbourne. Until von Mueller's death, Ellis often sent her paintings to him so he could classify and name the plants. Sometimes she included dried specimens as proof, even from as far afield as north Queensland.

It was probably while accompanying her husband on his business trips around Australia that Ellis met the English flower painter and world traveller, Marianne North (1830-1890), at Albany in Western Australia in 1880. This eccentric woman with a strong sense of vocation provided a role model for Ellis who later wrote:

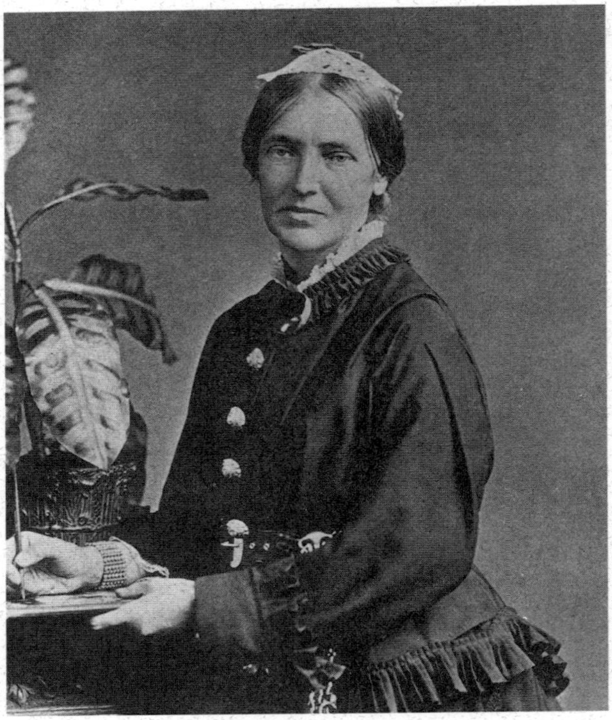

Ellis Rowan's mentor — the English flower painter and world traveller Marianne North. (Courtesy of the Royal Botanic Gardens, Kew)

> I became her devoted admirer, and she became the pioneer of my ambition. A world-wide traveller in search of specimens, her description of her adventures was so vivid, so graphic, so thrilling in its prospects of wider fields that I became infected, stimulated by an example and a result beyond dreams successful. Retiring to my room that night, after my conversation with her, I resolved to do as she had done. I would travel the world in search of flowers rare and wonderful, travel countries inaccessible, as well as those which offered difficulties only imaginary.[8]

Like her older mentor, Ellis stressed the importance of painting from life to show the flowers in their native habitats, and likewise suggested the habitats in her painting, especially after she took lessons in landscape painting from the Melbourne artist John Mather some time between 1905-08.[9] Again like Marianne North, Ellis wrote lucid accounts of her travels, published later in serial and book form, and was determined to place her paintings in the public domain. Miss North had the means to present her collection of paintings to the Royal Botanic Gardens at Kew in England, together with a fine gallery to house them. Ellis, by contrast, had to do battle with Australian state and federal politicians to sell her paintings for considerable prices. Ironically, Ellis was the better painter.

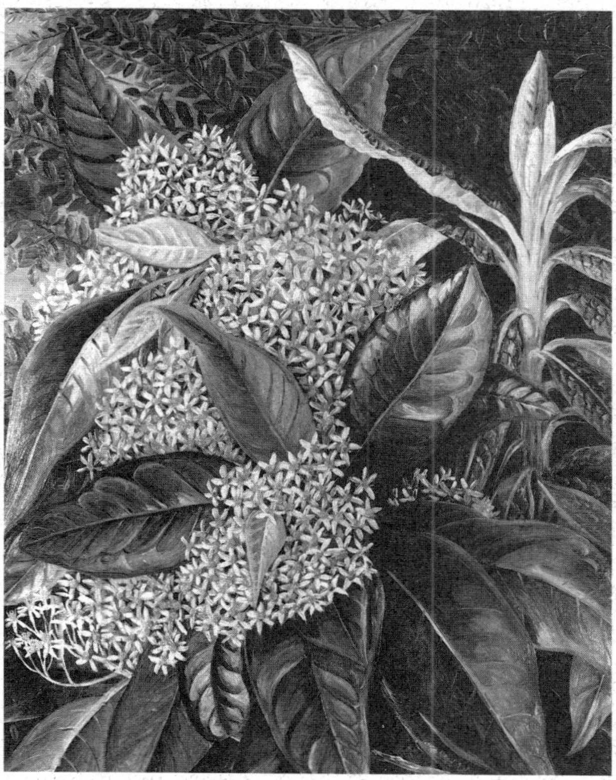

Marianne North's painting of Olearia argophylla (Labill.) F. Muell., Musk Tree and background of Nothofagus cunninghamii, (J.D. Hook.) Oersted. Evergreen "Beech", painted in Victoria during her Australian visit of 1880-81. (Marianne North Collection, courtesy of the Royal Botanic Gardens, Kew)

By 1887 when Ellis arrived in Queensland she had made the daring transition — quite unheard of in the Victorian era — from wife and society hostess to world-travelling artist, having already visited India and Europe in 1883-84. As her business partner Frederic Rowan gave his blessing to his wife's travels, and on 19 August 1887 on the eve of her depature for Queensland he wrote to the Premier Sir Samuel Griffith seeking official introductions to the Colonial Botanist and the curator of the local botanical gardens.[10] The real victim of Ellis's dogged pursuit of her own career and success was her son Eric, a wayward youth who was brought up mostly by the Ryan relatives and who died in South Africa at the age of 22.

In following years Ellis was to make repeated flower-hunting expeditions to Queensland. She invented a socially acceptable excuse for her long absences from her husband and son during the colder months: her delicate constitution could not withstand the severity of Melbourne's winters.[11] A tiny wasp-waisted figure with an "ultra-sweet voice" [12] and dressed like a lady of fashion even in the tropics,[13] her appearance belied enormous physical stamina and "the pluck and endurance of an explorer".[14] Her account in *Flower-Hunter* bears testimony to the miraculous improvement in her health once she reached Queensland — improvement enough to endure five hours' non-stop climbing of Mount Macmillan when she visited the Bloomfield River.[15]

Ellis came to Queensland armed with letters of introduction to notable colonists who provided her with a network of hospitality and support. The doors of privilege, and even officialdom, opened wherever she went. Her niece the late Lady Casey wrote that Ellis, witty, articulate, obstinate and calculating, had two effects on her contemporaries:

> Either they were mad about her and forever her blinded slaves, men and women alike, or they found her altogether too odd and were frightened of her.[16]

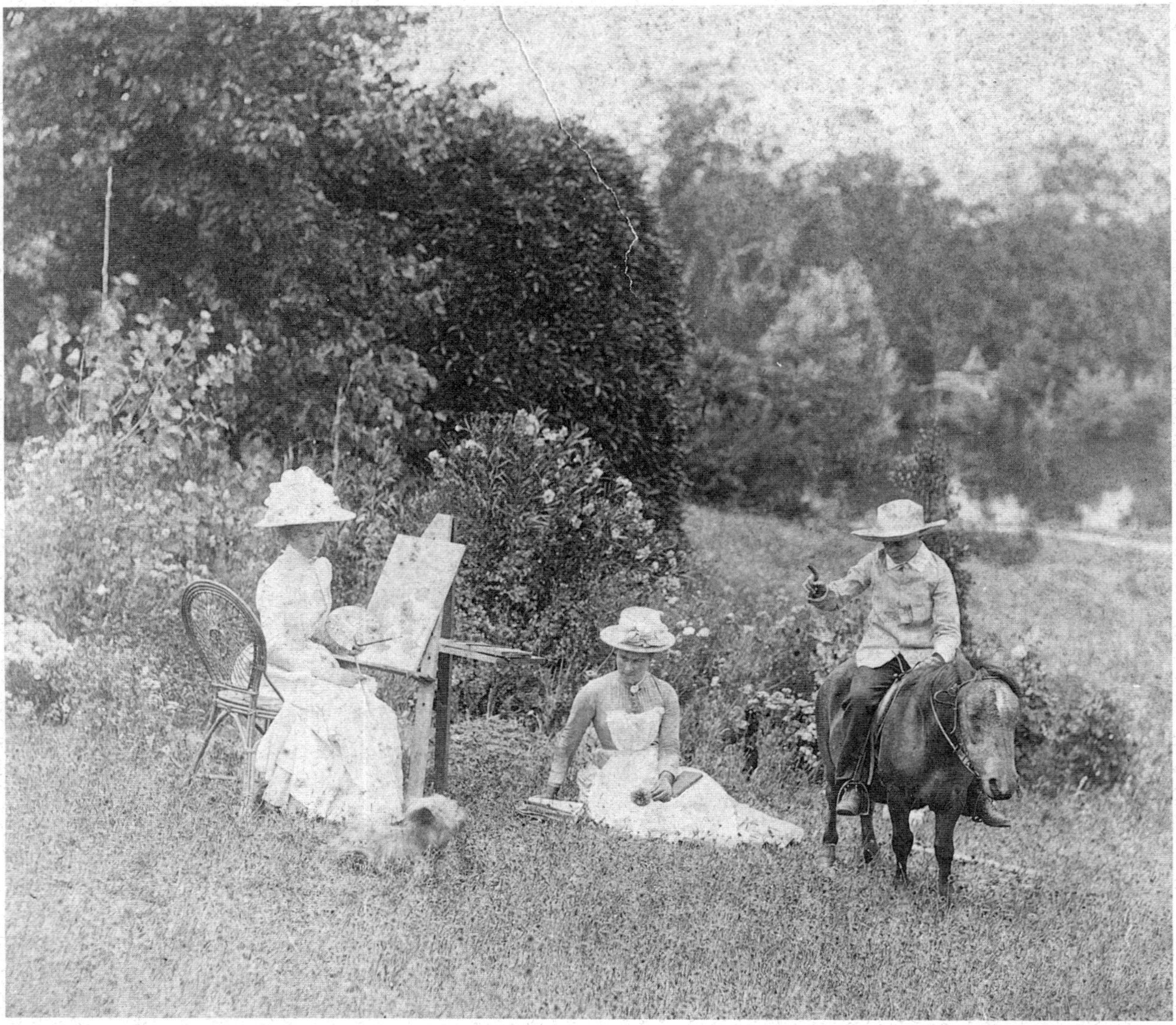

Ellis (left) *with her son Eric and Mary Moule (later Mrs Alexander Leeper), 1887.*
(National Library of Australia, by kind permission of the Misses Leeper)

Ellis first visited Queensland with the purpose of extending the range of illustrations for the Picturesque Atlas Publishing Company's proposed publication,[17] but was so fascinated with its floral wealth that she returned again and again.[18] Looking out over the Bloomfield River valley she exclaimed:

> ... if our Australian artists only knew what rich and endless subjects they would find in Northern Queensland, they would surely make up their minds to endure a little roughing and camping out, and take three or four months' holiday at this time of year. It would well repay them.[19]

Ellis was fortunate enough to travel when much of Queensland was botanically unexplored and its natural beauty unspoilt, when she could share in the excitement of finding rare or even unknown species.[20] She summed up the virtues of the colony:

> Queensland has the attraction for an artist that if one misses the wild flowers there are the wild fruits to follow, vividly coloured, as is everything else in that land of sunshine. And through the scrub, where the wild fruits are thickest, hordes of flying foxes sweep, and fruit-eating pigeons in all the colours of the rainbows ... North Australia in flower time is a blaze of colour.

And the vices:

> But there are risks and discomforts innumerable as well as pleasures to the artist who goes flower hunting in Queensland ... venomous snakes ... the nettle tree which, whether it be grasped gently or firmly, leaves hands swollen and stinging for days.[21]

Following Frederic Rowan's sudden death from pneumonia on 11 December 1892,[22] Ellis adopted half-mourning dress — a widow's bonnet — and resumed her travels, back to New Zealand in 1893-94 and then to London in 1895-96. Through the marriage of her sister Ada to an English aristocrat, Lord Charles Scott, she had an entré to London society and later convinced at least one journalist that she too held an aristocratic title![23] She managed to present three paintings to the ageing Queen Victoria, who had them made into a screen for her own room.[24] At her exhibition at the fashionable Dowdeswell Galleries in April 1896 Ellis introduced London society to the beauty of Australasian flora, and made a handsome profit in the process after Princess Mary, Duchess of Teck (the mother of Queen Mary), "led off" by buying two Queensland paintings.[25]

Besides selling well, the paintings received critical acclaim, notably from Lord Leighton, one-time President of the Royal Academy of Art. Ellis proudly published the exhibition reviews as a booklet, *Press Opinions of Mrs F.C. Rowan's Watercolour Flower Drawings*, printed later in London. While there she tried her skill at decorative painting with a series of panels of "the best among the varied flowers found in Queensland" for Lord Newton's townhouse in Belgrave Square and sought — in vain — similar commissions for Windsor Castle[26] and for the Queensland Court of the recently opened Imperial Institute.[27] Before leaving Europe she took her paintings to Germany where, she claimed, she met the former Prime Minister Otto von Bismarck and was offered £15,000 for her collection.

From 1897 Ellis travelled in the West Indies and in the United States, illustrating three text books on wild flowers for the young American botanist Alice Lounsberry (1872-1949) and exhibiting her paintings at New York and elsewhere. While in America Ellis became one of the first women to undergo a face-lift which, together with her "unreal" red hair, was a desperate attempt to preserve her youthful appearance.[28]

International recognition behind her, Ellis returned to Australia in 1904 to pursue her goal of recording native flora. She made her family's cottage at Mount Macedon her home base and financed her expeditions to the various states with regular exhibitions. In 1907 South Australia became the first State to purchase a large

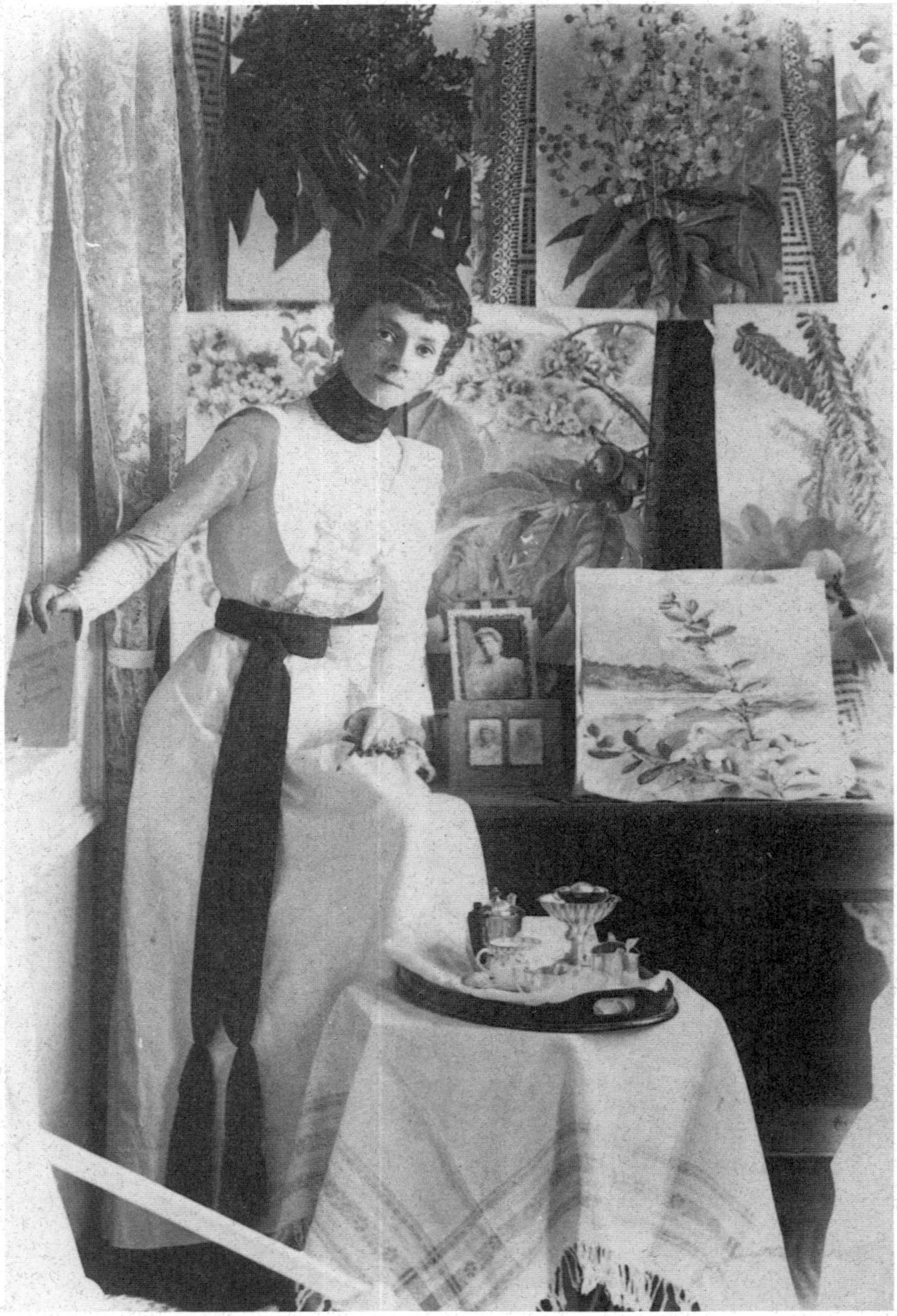

Mrs Rowan takes tea, c.1901. The naturalist Charles Barrett wrote of a meeting with her: "It seemed strange that this charming woman should speak calmly, over the tea cups, of dangers and difficulties encountered in travel in far countries, and the remotest parts of Australia". (Ellis Rowan Cuttings Book, National Library of Australia)

collection of her paintings, followed by New South Wales in 1910 and Queensland in 1912. In 1908 Ellis published a charming children's story about her pet bilby, Bill Baillie, adopted during her visit to the Kalgoorlie goldfields in Western Australia.

Prompted by a commission from the Sydney jeweller William H. Flavelle for a series of designs for china, Ellis made extensive visits to New Guinea in 1916 and 1917. By then she was in her late 60s. Travelling through rugged cannibal country to paint the endangered birds of paradise, she fell victim to malaria which broke her health. Her final success was to stage the largest exhibition of paintings then known in Australia, showing over 1000 works at Anthony Horden's Gallery, Sydney, in March 1920. According to the art critic William Moore, her takings of over £2000 set a national record for a female artist.[29]

But her male professional colleagues were to make another outcry when in 1921 the Federal Government decided to purchase a large collection of her paintings in response to a public campaign organised by the Rowan Collection Committee of influential supporters. Well-known artists protested that her work was quite unfit for any art gallery, let alone worth the asking price of £21,000.[30] Norman Lindsay, the master-painter of buxom flesh, had the audacity to call her work "vulgar" and to condemn the federal acquisition as "a deplorable commentary on public taste".[31] The Bruce-Page Government eventually offered £5000 for 947 paintings, paid a year after Ellis's death on 4 October 1922. The federal collection is now housed in the National Library of Australia.

QUEENSLAND

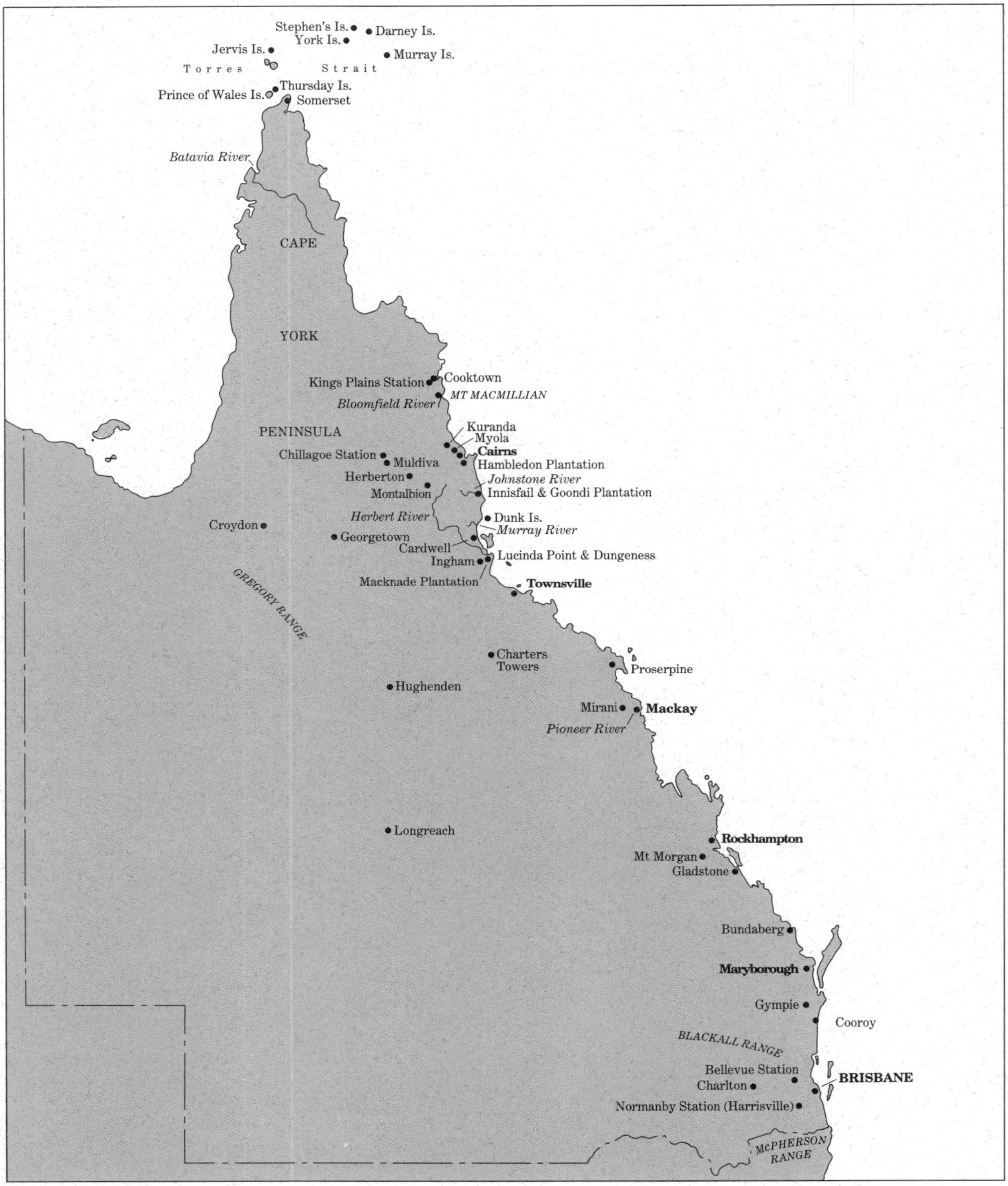
Stephen's Is.
Darney Is.
York Is.
Jervis Is.
Murray Is.
Torres Strait
Thursday Is.
Prince of Wales Is.
Somerset
Batavia River
CAPE
YORK
PENINSULA
Kings Plains Station
Cooktown
MT MACMILLIAN
Bloomfield River
Kuranda
Myola
Cairns
Chillagoe Station
Muldiva
Hambledon Plantation
Herberton
Johnstone River
Montalbion
Innisfail & Goondi Plantation
Croydon
Herbert River
Dunk Is.
Murray River
Georgetown
Cardwell
Lucinda Point & Dungeness
Ingham
Macknade Plantation
Townsville
GREGORY RANGE
Charters Towers
Proserpine
Hughenden
Mirani
Mackay
Pioneer River
Longreach
Rockhampton
Mt Morgan
Gladstone
Bundaberg
Maryborough
Gympie
Cooroy
BLACKALL RANGE
Bellevue Station
BRISBANE
Charlton
Normanby Station (Harrisville)
McPHERSON RANGE

FLOWER-HUNTING IN QUEENSLAND

1887

In 1887, at the age of 39, Ellis set out on an ambitious scheme to illustrate the flora of Queensland. Farewelling her husband Frederic in Sydney, she boarded the steamer *Barcoo* and, after a brief stopover in Brisbane, departed for the tropics on 27 August.[1] She was heading for Mackay where, her husband wrote, "we are told many beautiful flowers are to be found at this time of the year".[2]

At Mackay Ellis stayed with the Rawson family at their cattle run The Hollow on a picturesque bend of the Pioneer River.[3] The wide-verandahed house overlooked a sea of waterlilies and dense jungle beyond which she described in *Flower-Hunter*:

> My first walk in the wild tropical jungle I cannot forget. I entered, sketch-book in hand, by a narrow little pathway, probably made by an alligator. I kicked, as I thought, a grey stick aside — it was a snake, and quick as lightning it darted off, while I grew hot and cold by turns. There was such a death-like silence about me that I felt an intruder there, and the thick and tangled mass of rank vegetation completely hid the sun from sight. A few steps farther on I came to an opening, and below me lay a miniature lake, its water covered with large blue lilies floating amid their leaves on which the sun shone through a network of graceful palms. Scarlet, yellow-eyed dragonflies skimmed over its surface, while presently a great butterfly tremulously fluttered past, and the sunlight, catching the metallic lustre of its wings, changed them to every rainbow hue. The trees were clasped and linked together by delicate tendrils, and climbing ferns and huge caladiums covered the ground. It was a scene of wild, mysterious beauty, but in the distance there was the hum of a thousand gossamer-winged and hungry insects, and I hurried on with my sketch, for the mosquitoes had already found me out. Too much wrapped up in my work to turn round, I pushed twice aside from my cheek what I took to be a hanging tendril; but surely it moved too quickly — one wild jump and I was yards away! It was a long tree snake that had fastened its tail to a branch, and, curious to find out what manner of being this might be that had disturbed its solitude, was gracefully swaying backwards and forwards. This was the climax, and with a good deal of slipping and scrambling, I left the scene without a regret.[4]

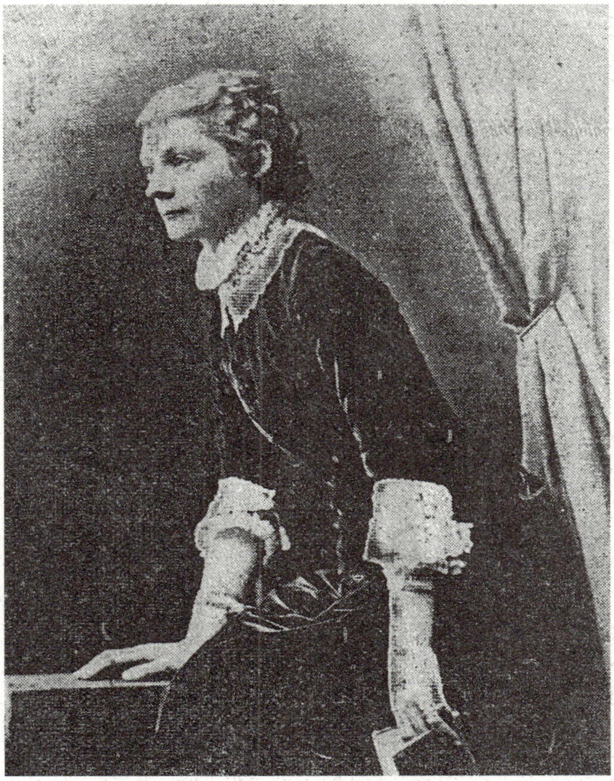

Ellis at about the time of her first visit to Queensland.
(*Australian Town and Country Journal*, 22 December 1888)

Indeed, Ellis concluded that "the snakes really *are* bad here" and painted the one that got into her hostess's aviary [5] (see page 63). From The Hollow, Mrs Rawson drove her guest north to Habana, a large sugar plantation worked by Kanakas and equipped with a modern mill. From Habana Ellis was able to make several excursions "to the different places of interest" in the district and found much to paint (cat. nos 61, 75, 91, 94, 96, 99). On one such excursion her host Edward Long climbed down a cliff face to bring her some orchids. [6]

From Mackay Ellis sailed north to Townsville where she stayed at the Queen's Hotel and enjoyed the gracious hospitality of the proprietress, Mrs Jessie Cran.[7] On 19 October she boarded the little trader *Palmer* which for many years linked Townsville with the sugar towns of the Herbert and Johnstone Rivers. Her first stop was the Herbert River. [8] Landing at the port of Dungeness, she travelled up river to Macknade sugar plantation as guest of the resident owners, Frank and Louisa Neame. During her six weeks' stay there she attended the annual show and races at Ingham [9] and painted cat. nos 57 and 68. Her output of paintings, by then at least fifty, was set back by an attack of malaria. [10]

After what she later described as a perilous, early morning journey downriver during a tropical storm — in Frank Neame's "cockle-shell" of a boat[11] — Ellis sailed on to Geraldton (now Innisfail) on the Johnstone River. From there she made her way upstream to Goondi sugar plan-tation where she stayed with the manager Charles Adams and his wife. Ellis marvelled at the jungle along the way:

> The river is very beautiful, and the thick, tangled vegetation comes down to the water's edge. It is so dense that it would be impossible to get through it except by cutting your way. From the tall palms above, trailing creepers crossed and intertwined their graceful plumes among fan-like branches, great bunches of scarlet berries drooped from a dark-leafed tree to the ground in heavy plumes. Here and there a bean tree (native chestnut) thrust a crimson-flowered branch through the dense foliage to catch the sunlight from above; large water hibiscus trees were everywhere shedding their blossoms, and the eddy of the current swept the crimson heaps in dozens under the banks, while ipomoeas, white, purple, blue and pink, wild ginger and endless trailing plants formed a thick undergrowth everywhere below. [12]

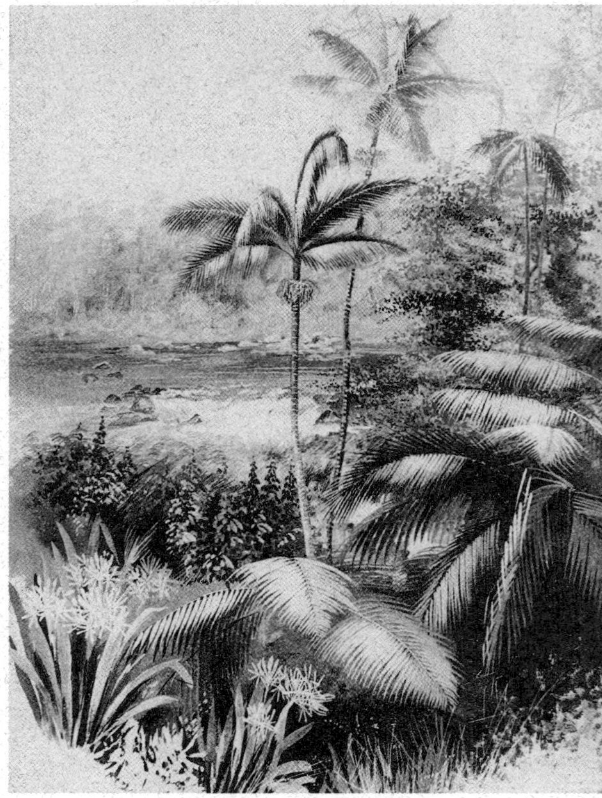

Ellis Rowan's painting of palms on the Pioneer River. (From an *Album of Photographs of Pictures by Ellis Rowan, presented to the Earl of Jersey,* Mitchell Library)

Goondi was a huge plantation worked, Ellis wrote, by no less than "700 souls", mostly Chinese. Her style was rather cramped at Goondi — she was unable to walk out into the jungle alone for

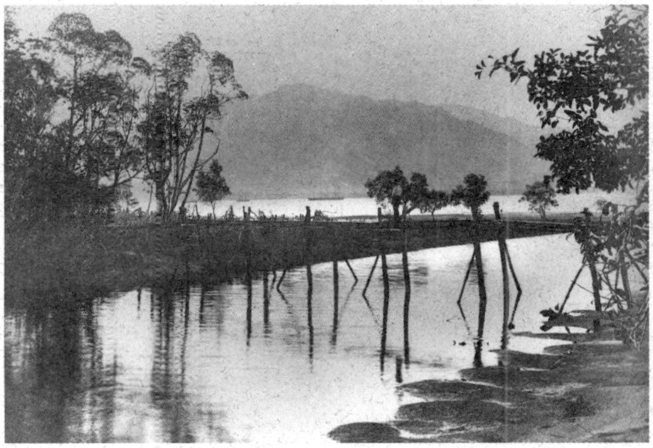

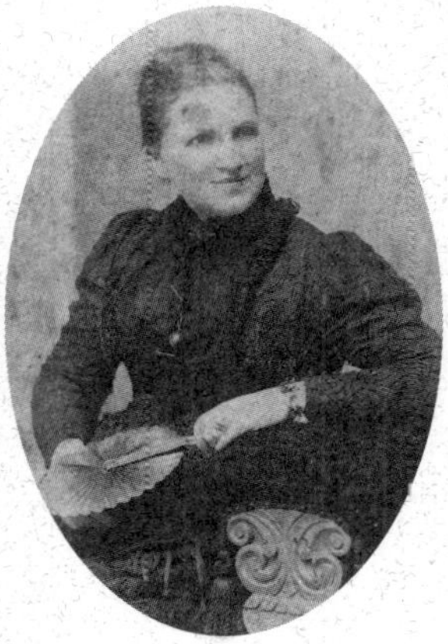

Mrs Jessie Cran, proprietress of the Queen's Hotel, Townsville, where Ellis first stayed in 1887 and again in 1891 and probably 1892. Mrs Cran filled Ellis's room with flowers to paint and served her "tea and real Scotch scones". (Courtesy of Mrs W.D. McKenzie)

Above: *The port of Dungeness at the mouth of the Herbert River, c.1890. Ellis called it "a miserable low-lying, dead-and-alive place" where she "broiled in the sun for five hours" waiting for the tide.* (John Oxley Library, 70090)

Below: *Goondi mill and plantation on the Johnstone River, c.1900.* (John Oxley Library, 24020)

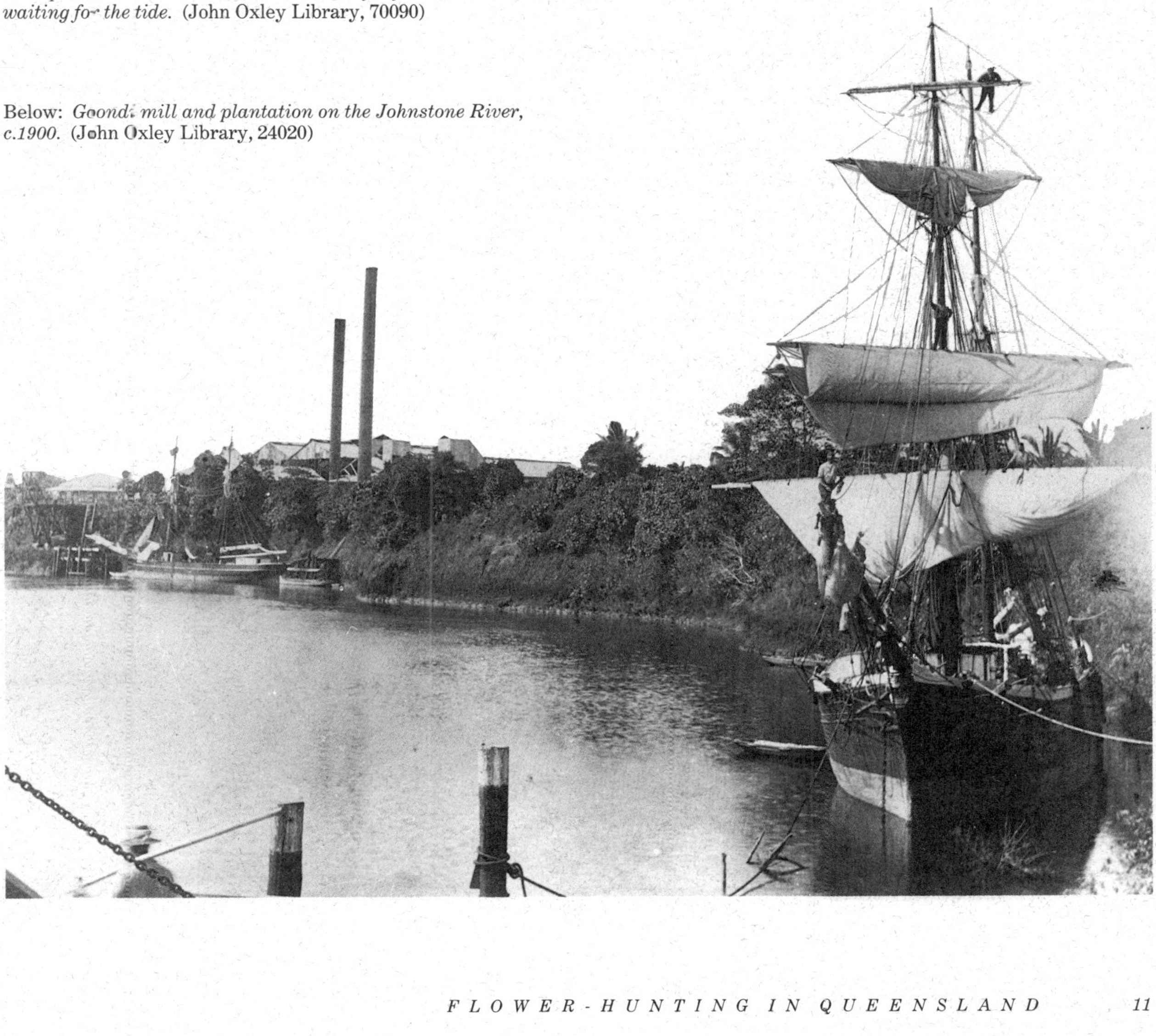

fear of attacks by the local Aborigines. Though disappointed to find that she was generally "too late for the flowers" (cat. no. 60), the fruits and berries were "quite worth painting" (cat. nos 66, 71, 97, 111). When she had another attack of malaria and the steamy summer heat began to turn her paints "quite liquid", Ellis decided it was time to retreat from the tropics. [13] Armed with a portfolio of at least sixty-four paintings, she arrived back at Townsville on 24 December [14] and returned quickly to the south. [15]

Back in Melbourne early in 1888 Ellis prepared her Queensland paintings for the Centennial International Exhibition opening in August that year. At this grand spectacle in Melbourne's Exhibition Building, which drew entries and visitors from all over the world, Ellis exhibited twenty flower paintings in the Queensland Court, [16] as well as several sketches of Queensland scenery and other paintings in the Victorian and New Zealand Courts. [17] Her contribution to the Queensland Court seems to have been a last-minute addition, recorded only in the addenda of the official catalogue. [18] Displayed along with the products and manufactures of the colony, the paintings were judged the most attractive of its fine arts exhibits [19] and won official acclaim. [20]

Another successful exhibitor was the Colonial Botanist F. Manson Bailey, who had assembled a collection of timbers for the Queensland Court and a collection of economic plants for the Queensland Conservatory. Bailey accompanied his exhibits to Melbourne and was one of the Queensland Commissioners for the Exhibition. If they had not already met in the north, Ellis would certainly have met Bailey at this event. He was to prove a useful contact in her future Queensland work.

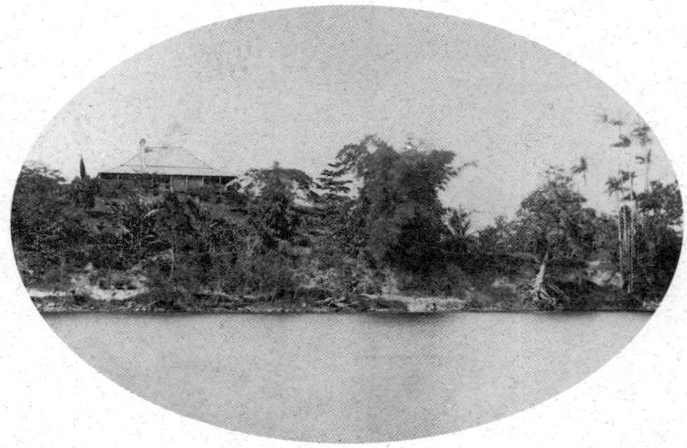

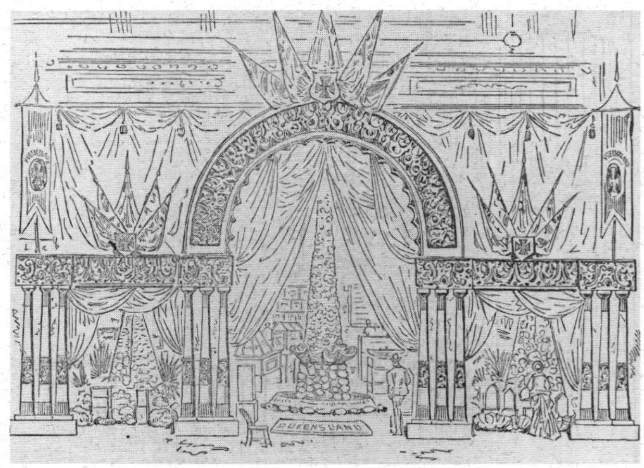

Above: *Macknade House — "a perfect one for this climate, built high up on the banks of the river, with a very fine garden all round it".* (John Oxley Library, 42514)

Below: *The grand "Mooresque" facade of the Queensland Court at the Centennial International Exhibition, Melbourne, in 1888 where Ellis exhibited more than twenty paintings and sketches. Pictures and maps were hung on the far wall behind the central trophy of shells.* (*Argus Exhibition Supplement,* 25 September 1888)

1891

Early in the winter of 1891 Ellis returned to Queensland for a more extensive visit north to Torres Strait, having travelled to Western Australia in 1889. She planned her northern journey at Normanby Station, Harrisville, where she and her son Eric stayed for a month with R. G. Casey Senior and his young wife Evelyn Jane.[21] Casey was an old acquaintance, having been a partner with Ellis's father in a stock and station agency in Melbourne.[22] At Normanby she painted in the extensive garden which, in the Caseys' time, reached down to Warrill Creek (cat. no. 74).

In July she moved on to Fernberg (now Government House) in the Brisbane suburb of Bardon, as the guest of Mr and Mrs John Stevenson, MLA,[23] friends of the Caseys. Ellis was almost certainly the guest of honour on 20 July when Mrs Jane Stevenson entertained Brisbane society at an "At home".[24] On her last afternoon in Brisbane Ellis visited the display of native and exotic plants at the Acclimatisation Gardens at Bowen Park.[25] William Soutter,[26] the energetic curator of the gardens and honorary secretary of the Queensland Acclimatisation Society, gave her an "armful" of tropical flowers.

Ellis sent Eric back to Sydney by train before boarding the steamer *Maranoa* on 1 August, together with R. G. Casey Senior who was on his way to his other pastoral properties in central Queensland.[27] On arrival at Rockhampton Ellis

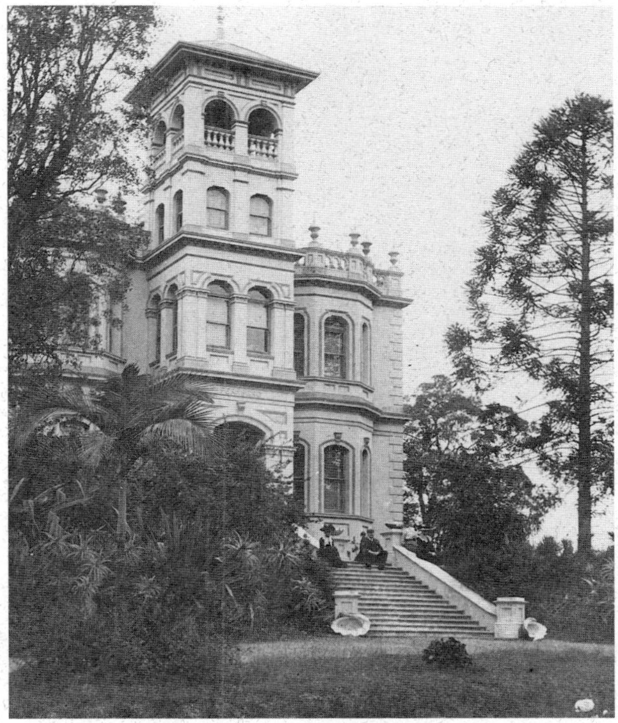

Fernberg in the time of wealthy grazier and company director John Stevenson, MLA, who added the elaborate facade.
(Courtesy of the Queensland Women's Historical Association)

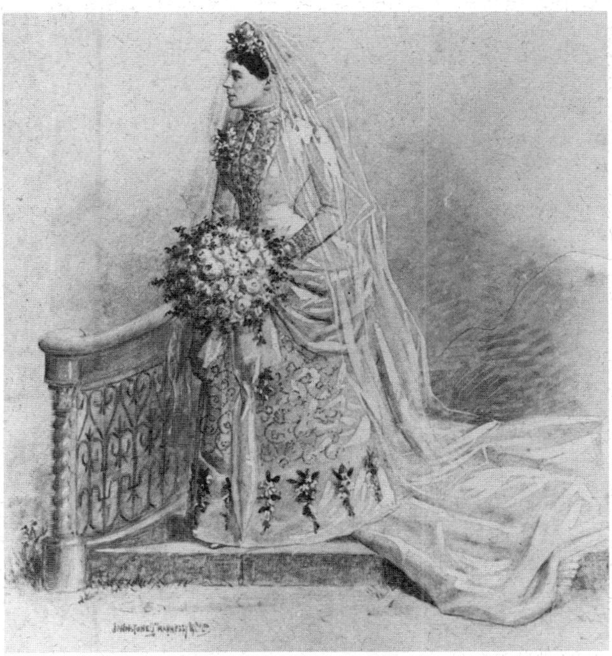

Evelyn Jane Harris at her marriage to R. G. Casey Senior in Brisbane in 1888. She and Ellis were long-time friends.
(Courtesy of the Queensland Women's Historical Association)

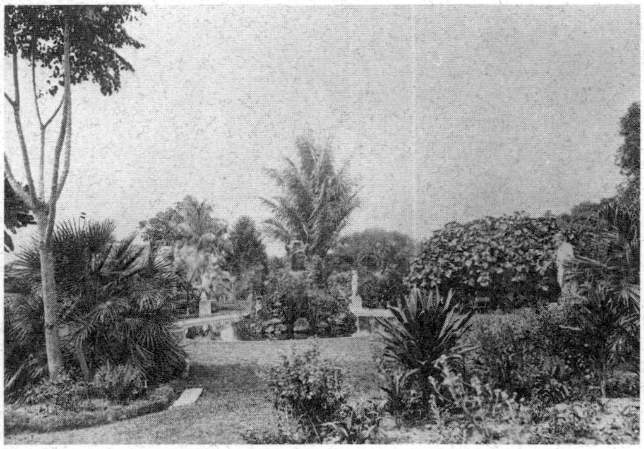

The Acclimatisation Gardens at Bowen Park, Brisbane, c.1889. (John Oxley Library, 100667)

William Saville-Kent's method of photographing submerged corals and bêche-de-mer.
(W. Saville-Kent, *The Great Barrier Reef*, W.H. Allen, London, 1893.)

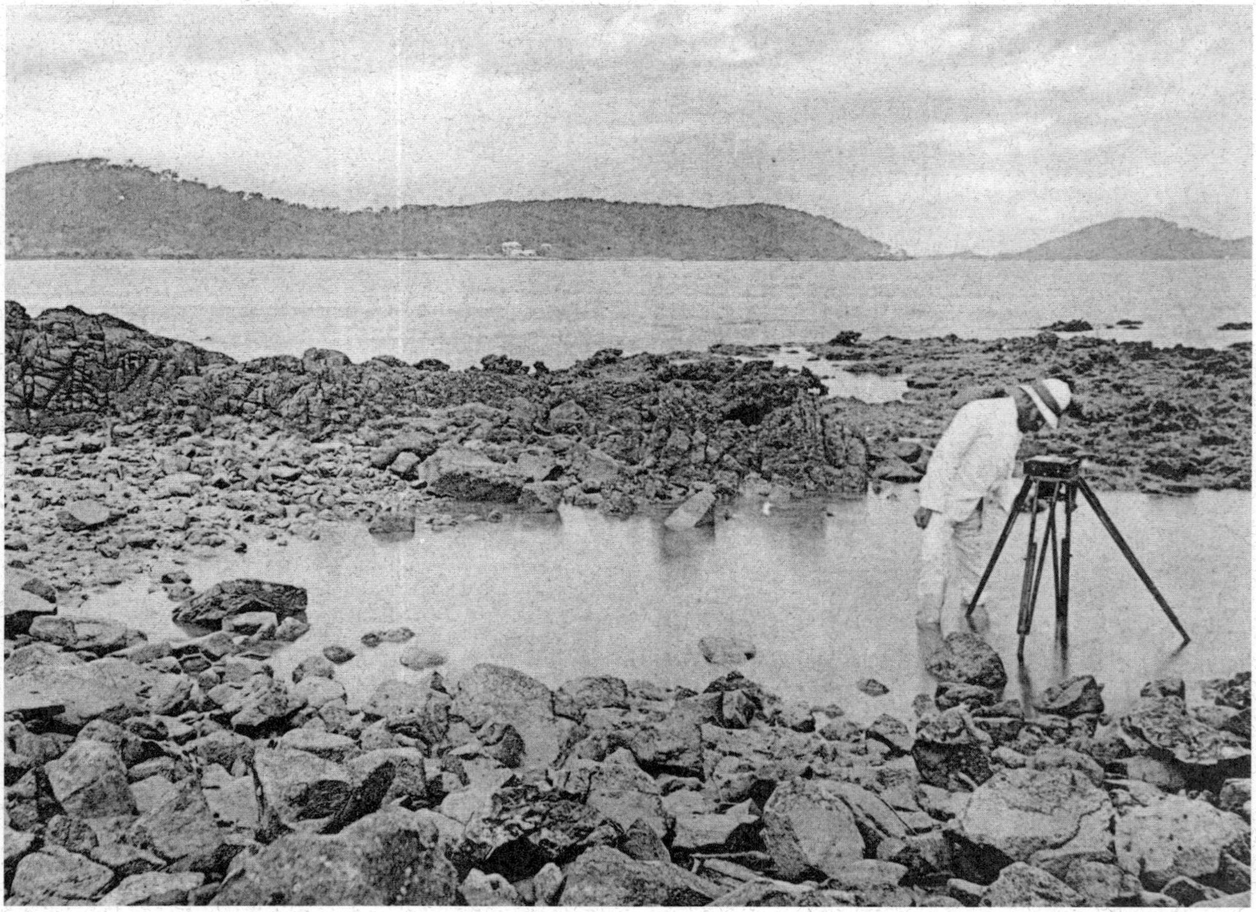

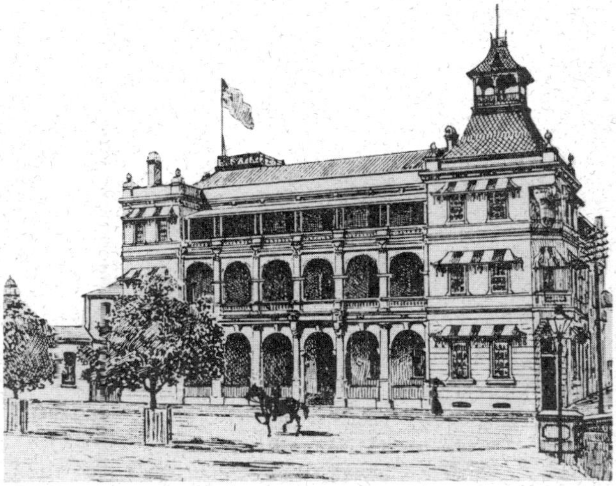

The Criterion Hotel, Rockhampton — "a capital hotel, prettily situated on the bank of the Fitzroy River". Ellis stayed there on her way northwards in 1891.
(John Oxley Library, 158488)

made a one-day visit to Mount Morgan with the wife of William Pattison, a major shareholder in the town's famous gold mine.[28] Ellis was herself a speculator in mining shares, and a keen observer of north Queensland's booming mining enterprises during her 19th century visits. After a week's stay at Rockhampton[29] she set a direct course for Thursday Island to see the spectacular visit of the Aust-ralian Squadron. Ellis was looking forward to meeting her brother-in-law Admiral Lord Charles Scott, on his last visit to the far north as commander of the Australian Station of the Royal Navy. On 11 August she trans-shipped at Townsville to the *Jumna* for Thursday Island.[30] There she was taken ashore in the steam launch of HMS *Orlando*, the Admiral's flagship, which had only recently arrived with Queensland's Governor Sir Henry Norman to join the eight warships anchored in the harbour.

Ellis registered at the Grand Hotel, from where she made early morning field trips to collect flowers "worthy to form a gift for our Queen"[31] (cat. no. 76). A fellow guest at the hotel, William Saville-Kent, Queensland's learned Commissioner of Fisheries,[32] also made early morning trips. At low tide he went to hunt and photograph on the island's reefs. The hotel verandah was littered with his colourful specimens of corals and sea anemones; off the verandah, Ellis's room was fragrant with her floral specimens. Ellis and Saville-Kent probably seemed a strange pair to the hotel's permanent residents, an assortment of screeching parrots and cockatoos which lived in cages on the verandah.

On 18 August Ellis accompanied the Admiral and the Governor to a fancy dress ball attended by "almost all" of the local residents.[33] Among the ball guests was one of the north's most celebrated pioneers, Frank Jardine,[34] and his charming Samoan wife Sana. They invited Ellis to Somerset, just south of the tip of Cape York, the seat of their prosperous cattle and pearling empire. From its establishment in 1863, Somerset had become a regular hunting ground for naturalists whom the Jardines entertained in style. Few ships passed by the well-known homestead without dipping their flags in salute. Ellis was stunned by this "new

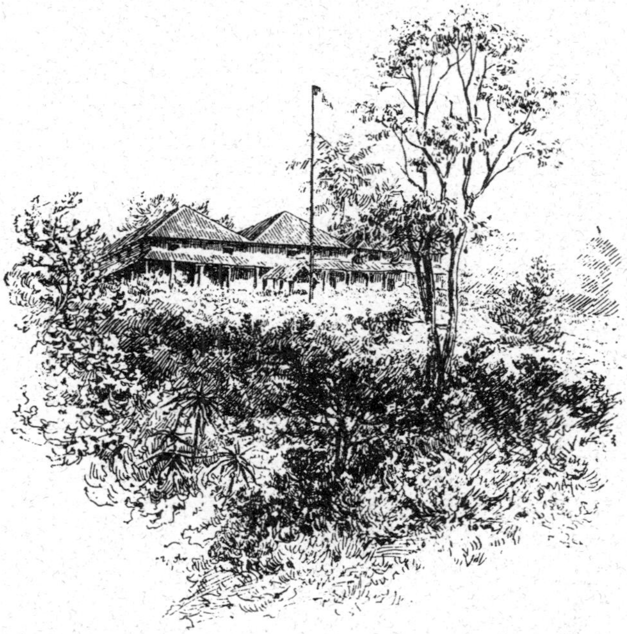

Ellis Rowan's drawing of the Jardine house at Somerset — "like other Queensland houses, is built on piles with a wide verandah all round, and the long and lofty rooms opening one into the other". (Flower-Hunter)

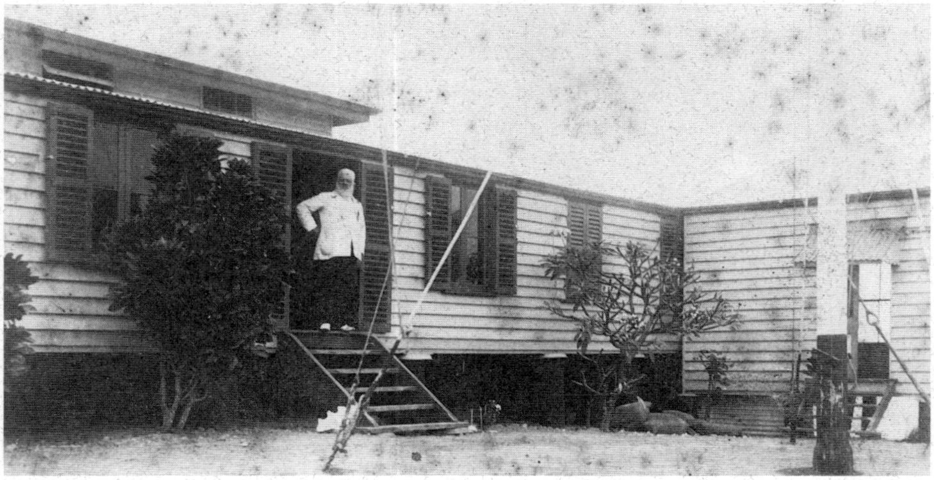

John Douglas at the Government Residency, Thursday Island, 1897.
(John Oxley Library, 49817)

Walton's Railway Hotel, Cairns, "conveniently situated opposite the Railway Station", c. 1891. Ellis was driven from the wharf in the hotel's cab seen in the foreground.
(John Oxley Library, 70142)

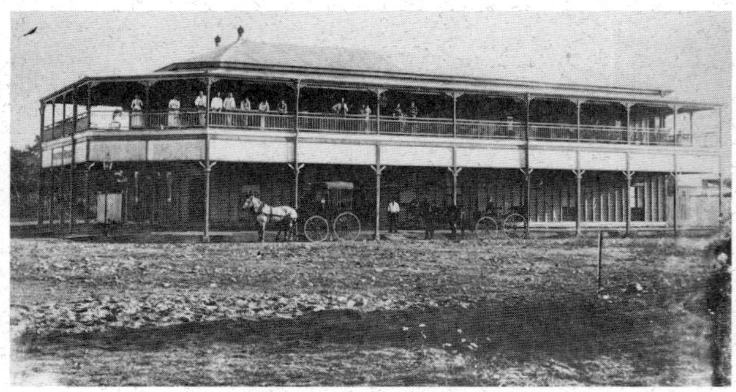

Eden",[35] which she found even more beautiful than the Johnstone River, and painted frantically (cat. nos 41, 46, 51, 59, 63, 77, 89, 93). She travelled south to the Batavia River[36] — this is probably the journey Ellis later described in newspaper articles as a "long ride" with Mrs Jardine, claiming they had a close encounter with hostile Aborigines.[37] At the end of her several weeks' stay at Somerset Ellis wrote:

> The too short day is over here, before I realise it has commenced. I wish I had the privilege granted to Joshua of power to arrest the sun. I never want to go to bed, and I grudge the night hours that I waste. I am up at half-past six; when I have a big tumbler of milk, and commence my painting; otherwise I could do nothing, and even now I must leave dozens of flowers to wither. I work on till the cool of the evening, when kind hands take my painting from me, and we go for a scramble, or the horses are brought round and we ride. We spend the evenings in our lounge chairs on the verandah, and the children, who have beautiful voices, sing to us.[38]

On 19 September[39] Ellis set off with Mrs Jardine on a grand tour of the islands of Torres Strait as far north as Jervis Island[40] (cat. nos 72, 84), hosted by the Government Resident John Douglas in the Government steamer *Albatross.* His hospitality became the centre of a public controversy later that month when a newspaper report from Thursday Island complained that the murder of some Aborigines at the Seven Rivers had not been officially investigated "because the *Albatross* was absent escorting Mrs Rowan, the artist, around the islands".[41] The issue of the "pleasure expedition" was taken up in Parliament by the Member for Ipswich, A. H. Barlow, who sought an inquiry into the Government Resident's conduct.[42] The Government, however, expressed confidence in Douglas, for he was a conscientious and able administrator who had held a long political career before his retirement to Thursday Island.[43] Douglas invited Ellis to stay at the Government Residency and was "untiring in his efforts to get her flowers".[44]

Ellis, according to a Melbourne newspaper

report, was well pleased with her visit to the far north — both with the "beneficial effect" on her delicate constitution and with the "excellent work" she had done there.[45] She sent the paintings she had done so far to Melbourne to be shown in an exhibition opening at Buxton's Art Gallery at the end of October, while she remained in the north "working, as usual, while there is an hour of daylight". The exhibition of nearly 120 works comprised Western Australian as well as Queensland paintings.[46]

From Thursday Island Ellis sailed to Cooktown[47] for a week's stay:

> ... but it blew such a gale there the whole time that, excepting for a few walks round about the town and to visit a native camp, I went nowhere.[48]

Her hostess there was Mrs Anne Rebecca Bauer, widow of a German sugar planter, Friedrich Bauer, and a family friend of Baron von Mueller. Ellis sailed on to Cairns, arriving on the *Aramac* on 3 October.[49] She stayed at Walton's Railway Hotel, where she was so comfortable that she later decided to stay at its sister hotel at the mountain resort of Myola, then the terminus of the Cairns Railway.[50] But before her excursion on the railway she spent a "lazy fortnight" at Hambledon sugar plantation on the foothills of the Bellenden Ker Ranges to the south of Cairns — "one of the most beautiful places I have seen in Queensland".[51]

The climax of Ellis's visit to Cairns was a thrilling ride on the railway up the Cairns Ranges, travelling, so she claimed in *Flower-Hunter*, on the cowcatcher of the engine:

> I have now travelled ... on the cowcatcher of an express engine ... I wouldn't have missed it as a novel experience, but have no wish to try it again ... The line by which I travelled has only been opened for four months, and it is by far the most beautiful, they say, in Australia. Besides being a good specimen of engineering skill in

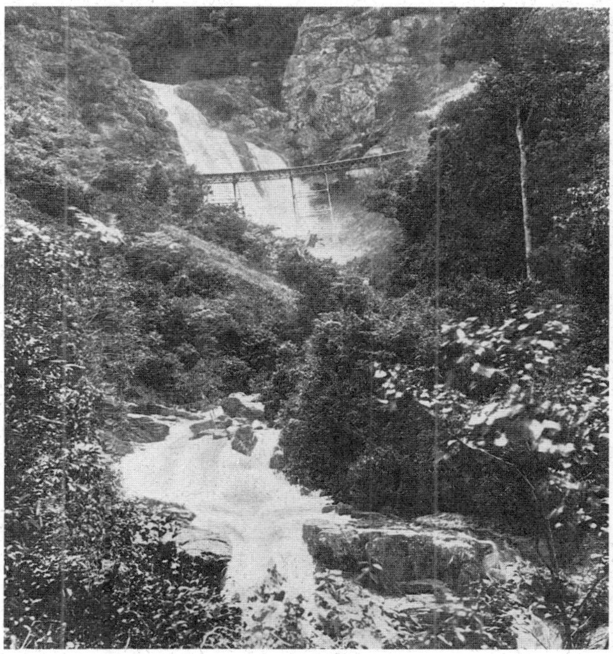

Stoney Creek Falls on the Cairns Railway, c.1900. (John Oxley Library, 143736)

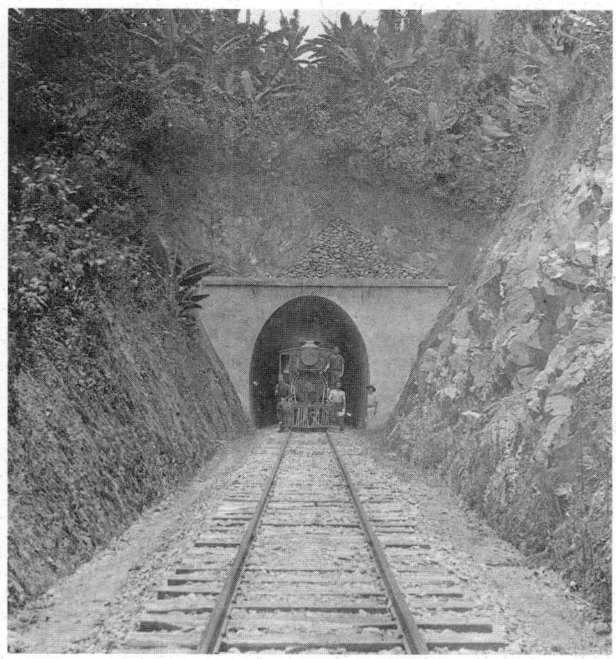

On the Cairns Railway — Ellis claimed to have ridden on the cowcatcher of a train. (John Oxley Library, 27394)

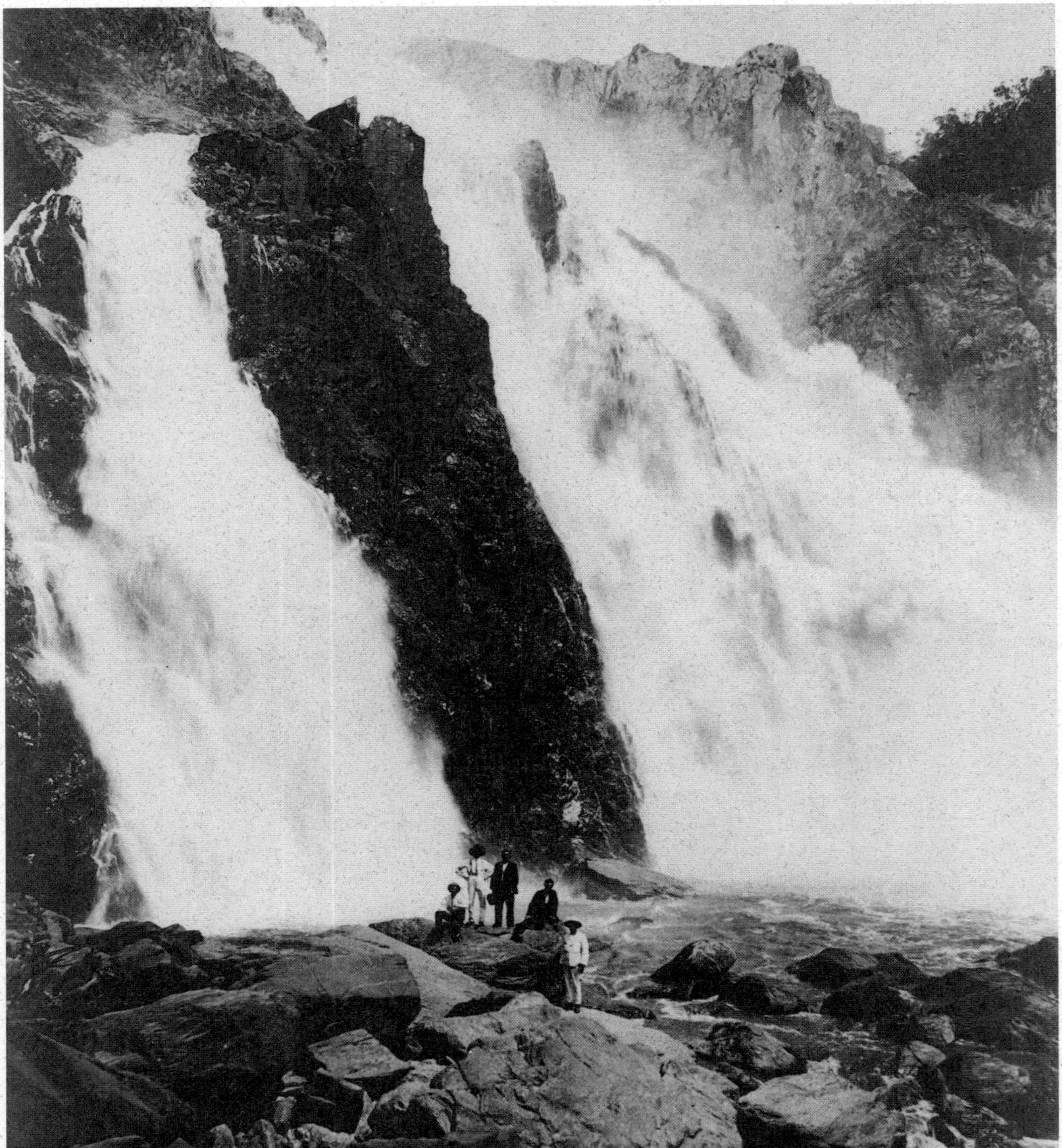

The Barron Falls — "Queensland's Niagara" — the north's premier tourist attraction at the turn of the century.
(John Oxley Library, 158123)

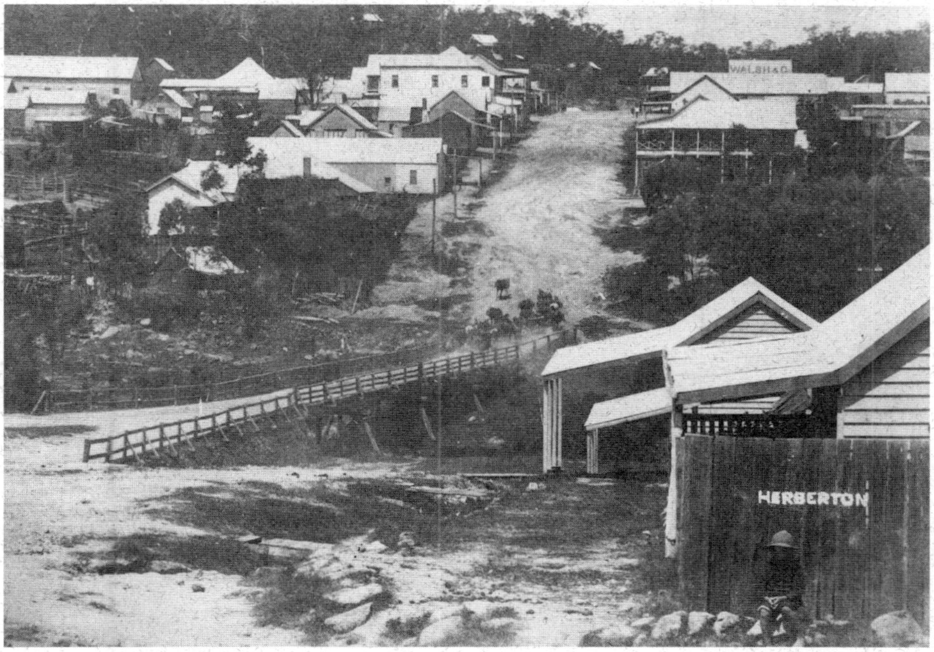

Herberton, c.1895, soon after Ellis Rowan's visit. At Herberton she had her first experience of an "up country" inn — a daunting experience for any genteel guest. (John Oxley Library, 62161)

mountain railroad making, the beauty of each fresh turn becomes more and more impressive as it mounts the higher. In many places the train seems almost to hang over the precipices 600 and 900 feet deep, while still towering above are masses of granite rock and of wild tropical vegetation, sometimes palms, vines, bananas, or fairy draperies of hanging ferns; sometimes clusters of orchids, with scarlet flame-trees here and there, giving light and brilliant colour to the dense masses of vivid greens, which range from the palest apple and russet to the deepest olive.

There are waterfalls in every direction, and presently below, far down in the depths of vegetation, one sees the Barron River, and away in the distance, between the mountains, like another world below, the valley dotted with houses, with the blue sea beyond and the faint outlines of other distant mountains and the little indistinct town of Cairns almost buried amongst them. One longs for a few minutes to enjoy it all, but the merciless train keeps on, while here and there below one catches glimpses of the river dashing over great rocks; or of passing gleams of white foam at the foot of the falls, and finally, for one speechless moment, the falls themselves. I could only hold my breath and gasp "Oh!" for I was on the engine, but the driver gave no response, excepting to heap more coals on the fire. Another second and the vision was gone, and we found the river high up beside us, breaking and eddying round great rocks, or else, without a ripple, gliding along over sandy bars.

Gradually the train slackened its pace and we were at our journey's end . . . [52]

If Ellis had been riding on the engine's cowcatcher, it is hard to imagine how the driver, shovelling coal, could possibly have heard her little gasp of astonishment. She repeated the story in 1905 for the *New Idea* women's magazine. Ellis claimed an urgent need to return to Cairns [53] had prompted her "memorable journey," but in the original version she had travelled in the opposite direction.

From Myola, which she reached by 20 October, [54] Ellis rode on a railway trolley down to the Barron Falls where she claimed she was lowered by rope down a cliff face to paint a rare orchid.[55] By 1912 a snake had become co-star of the story: "man's enemy, the snake, was watching her movements with disapproval" as she dangled from the rope.[56] More likely she went down a well-trodden tourist track beside the falls. One of the railway workers climbed a tall tree to bring Ellis "a beautiful yellow cassia that hung in drooping yellow sprays and then brown pods 14 inches long"[57] (cat. no. 22).

Ellis left Myola on 30 October[58] for a tour of the tourist attractions and mining enterprises in the Cairns hinterland, motivated as much by a lust for adventure and mining speculation as by a need to paint flowers. She took a Cobb and Co. Coach to Herberton, then on to the "celebrated" Muldiva silver mine[59] and to the Chillagoe Caves over "the roughest road in Australia"[60] — so rough that the journey beyond Montalbion had to be taken in an open buggy which received "the full benefit of the sunshine and dust".[61] She stayed at Chillagoe Station whose owner, William Atherton, guided her through the "Cimmerian darkness" of the caves.[62]

Ellis's *Letters from Queensland*, 1891, record that she had returned to her hotel at Myola by 5 November to enjoy ten days' rest following the "hideous" journey which had left her "burnt nearly black" and her arms and shoulders blistered with sunburn, even through her jacket.[63] Indeed, she wrote that she spent six weeks in all at "that little bush home" at Myola.[64] The *Letters* contradict her account in *Flower-Hunter* where she claimed that from Herberton she hired an open buggy and companions to undertake an extensive journey westwards to Georgetown to see and paint the wild flowers of the Gregory Range.[65] The journey to Georgetown was probably imaginary — or its account at least partly inspired by a series of travel stories about the "North and West" which had recently appeared in the widely read *Queenslander* newspaper and also in the *Brisbane Courier*.[66] It is remarkable that the collections of the Queensland Museum and the National Library contain no paintings from this supposed journey.

After a day's stopover at Cairns[67] Ellis again retreated from the tropics. She caught the steamer *Arawatta*, and arrived in Brisbane on 27 November.[68]

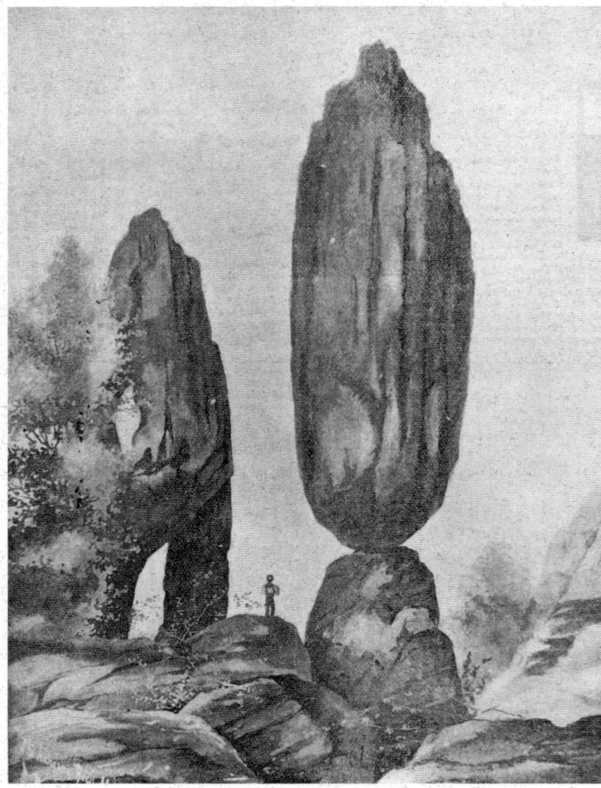

Ellis Rowan's impression of the hanging rock at Muldiva. (*New Idea*, 6 April 1905)

1892

Ellis returned to Queensland in the following winter. She headed straight for Cooktown on the *Arawatta*, arriving there early in June after brief stopovers at Brisbane and Townsville.[69] At Cooktown she met another notable visitor, a German biologist Richard Semon, who reported that Ellis was "staying there for some time" and intended to visit New Guinea.[70] Again she stayed on the outskirts of Cooktown with Mrs Bauer and her family who went to considerable trouble to get her flowers and accompanied her on "bush scrambles" around Cooktown[71] (cat. no. 67). Ellis taught one of Mrs Bauer's daughters to paint. Anniechen Bauer (later Mrs MacArthur) was a gifted pupil and became a well-known north Queensland flower painter.[72] She sent paintings to the Queensland International Exhibition of 1897 and to the Queensland preliminary exhibition for the Australian Exhibition of Women's Work of 1907.[73]

With the Bauers' help, Ellis planned a flower-hunting expedition to the remote Bloomfield River

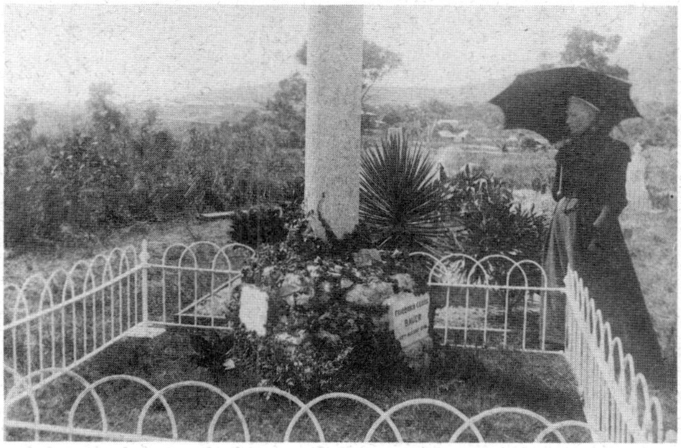

south of Cooktown, where Friedrich Bauer had been manager of the Vilele sugar plantation. There is no firm evidence to date the Bloomfield expedition to mid-1892 but, in view of Semon's comments, that date seems probable. Moreover, the Bloomfield expedition is not mentioned in Ellis's *Letters from Queensland*, 1891. On "21" (June?) she left Cooktown on the gunboat HMS *Lizard*, but her letter to her host had gone astray.

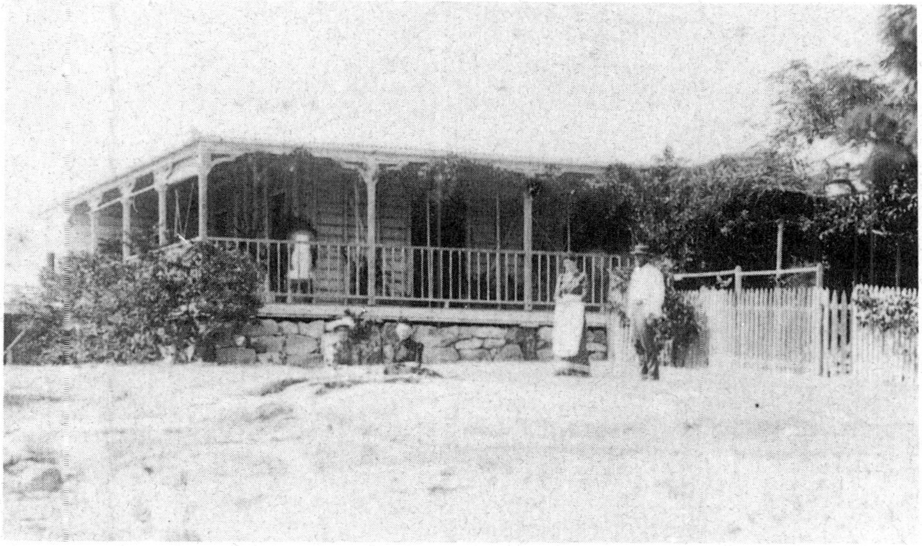

Above: Ellis Rowan's hostess Mrs Anne Bauer at the grave of her husband Friedrich who died at Cooktown in 1890. (Courtesy of Mr Peter Croker)

Left: Rubislaw, the Bauer residence in Cooktown, at about the time when Ellis Rowan stayed there in 1892. Mrs Bauer and daughters are in the foreground, while their Aboriginal maid Maggie stands on the verandah. (Courtesy of Mr Peter Croker)

She found herself disagreeably detained at the mouth of the river before she could make contact with her destination, Wyalla Station, further up the picturesque Bloomfield valley.[74] Becoming impatient, Ellis tried to cross the fast-flowing river by stepping on fallen trees, only to become stranded midstream. She recalled yet another moment of peril in *Flower-Hunter*:

> Between the root of the tree and bank there was now a great gap where the tide was rushing through with tremendous force, and close alongside of me there rose something that, for the moment, I thought was another half-sunken tree. Then it fell, a gray, loathsome creature that almost paralysed me with fear as I marked the long line of its greedy-looking jaws. I knew that the river teemed with alligators, but, somehow or other, I had never given them a thought. Its horny back was not more than a foot below me, and I hardly dared to breathe, much less to move. It slid along under the log and I felt the vibration of its body rubbing as it came up on the other side, then it turned with its head up stream again, its snout just above water as if it smelt game. Uncertain as to its movements, it sluggishly played round and round. My eyes were riveted on it, and in the horror of the moment I forgot the river, tide, and everything else, as with the rising water it came so close again that my feet almost touched it as it stirred the slimy ooze and mud from the bank with its tail. For a few seconds the voracious monster lay apparently insensible to everything, but with its ears open to the slightest sound; I hardly dared to draw breath. Now, as if waiting for the supreme moment, its opportunity to spring, it rose the full length of its body and menacingly clashed its jaws, then with snout down stream, it went under, leaving nothing in its wake but a long ripple on the surface of the water.[75]

She was eventually rescued by a party of Aborigines.

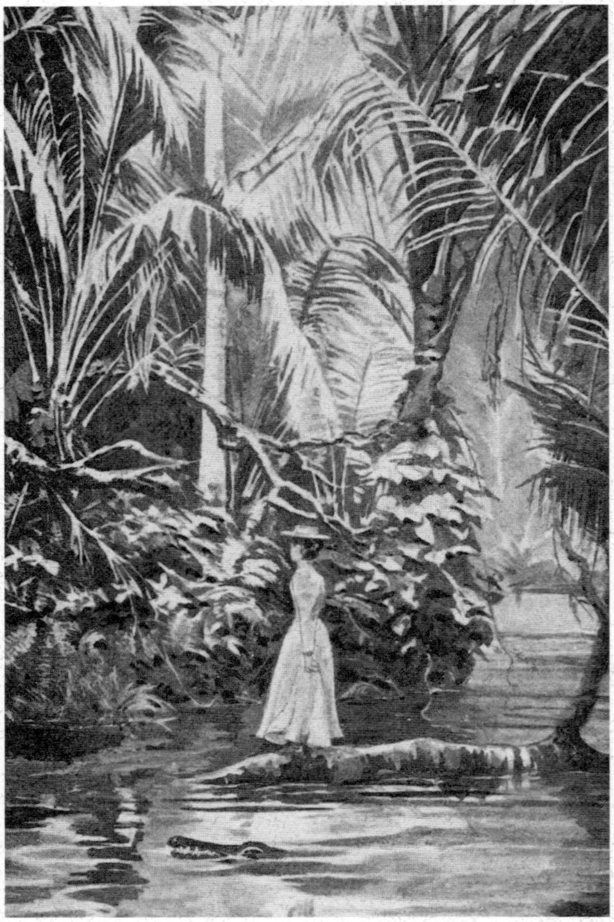

"Suspense!" Ellis Rowan's painting of her encounter with an "alligator" at the Bloomfield River. (*Flower-Hunter*)

Ellis was then taken upriver to Wyalla Station where she stayed for several weeks with George and Mrs Hislop in their "pretty bungalow house" near Mount Macmillan. With the Hislops she explored the ruins of Friedrich Bauer's former plantation, and visited a mission station and the "mighty" falls nearby. After a strenuous climb to the top of Mount Macmillan she looked out over the Bloomfield River valley to the coast:

> Thick clouds kept passing over us, but now and then through the sunshine we had an uninterrupted sight of perhaps one of the finest views in all Queensland.[76]

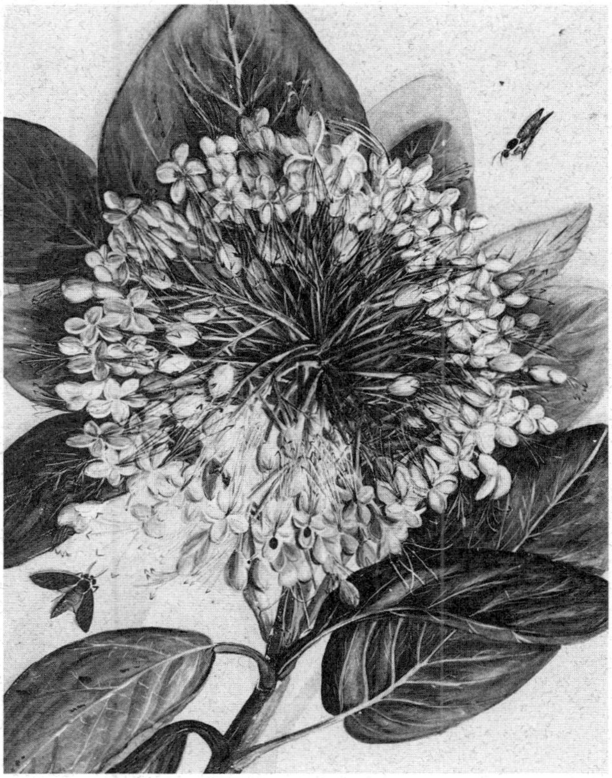

Anniechen Bauer's painting of Deplanchea tetraphylla (R.Br.) F. Muell., painted in 1892. (Courtesy of Dr and Mrs Drury Clarke)

The Bloomfield River valley, c.1884. Ellis called the deep valley, with its falls and rapids, "perhaps one of the finest views in all Queensland". (John Oxley Library, 103543)

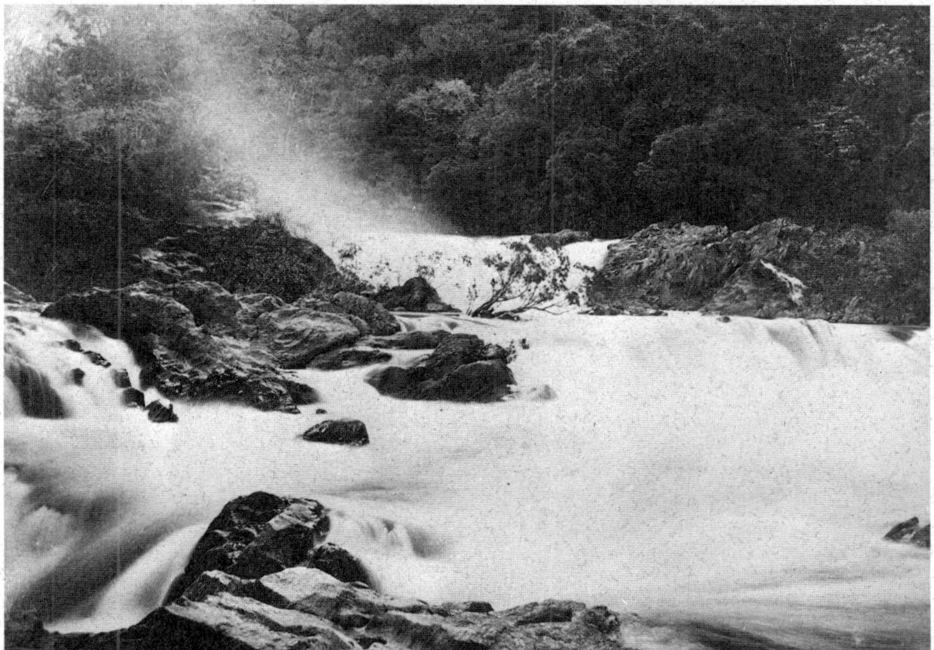

She found much to paint in the jungle beside the river "all hung and matted together with ferns and creeping vines" [77] (cat. nos 80, 82).

Ellis returned to Cooktown after her visit to the Bloomfield, but her movements over the next month or so are uncertain — possibly she went on her intended visit to New Guinea (cat. no. 42). According to a newspaper report she had certainly accomplished "a prolonged flower-painting expedition around the New Guinea coast and in the islands of the Torres Straits" before her return to the south later in the year. [78]

By late August Ellis had moved on to Somerset to set out on another "pleasure expedition" around the islands in the Government steamer *Albatross*. Again she was accompanied by Mrs Jardine, but this time her host was John Douglas's deputy, Charles Savage, the Chief Inspector of Police, who was on an official tour of duty. The steamer carried another famous pleasure-seeker, the artist Tom Roberts, who was on a three-month trip to the north. When they anchored at Murray Island, Ellis made quite an impression on Robert Bruce, one of the island's few white residents, whom Douglas had asked to entertain those "special people":

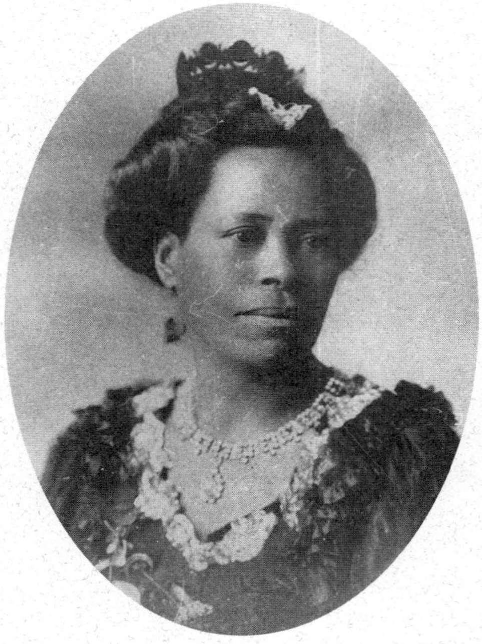

Mrs Sana Jardine, c. 1901. She was born Princess Sana Solia, niece of the King of Samoa. (John Oxley Library, 67272, by kind permission of Capt. G. L. Jardine-Vidgen)

> . . . Mrs Rowan I met before, just a year ago, at Mr Douglas's House in T. i. . . . She is a woman I fancy about 45 years old. . . She told me she had a son 18 years old yet she looks like a British girl 17 years of age — her skin is so pure, with just enough of blush colour to make her very pretty, of course she is slim built — I noticed that she at times, masauged her eyes & mouth with her fingertips that I have no doubt keeps up the circulation of the skin and keeps away wrinkles, of which she is very free — there were four gentlemen came out in the Steamer, all strangers to each other and to her, thay had very bad weather Mrs R. being sick all the time, yet when I went off to bring them ashore — she excused herself for being untidy as she had not had a wash nor a change since she came on board — yet she was the freshest of the party as far as looks went — This woman works hard, when at Douglases she was painting from daylight till dark, constant, except at Breakfast and tiffen. [79]

As at the Bloomfield, at Murray Island Ellis showed remarkable stamina for one of such delicate constitution. Bruce continued:

> . . . she done some hard walking here on the Sunday, climbed up Ferndale, right up the rocks holding on by the vines and creepers and would axcept help from no one, saying she could follow where any man would go — two much younger women Captain Reid's wife and Sanna (Mrs Jardine) who was acting as Mrs R's maid required help at many stiff places — Captain Reid proposed after we had got to the top of the gully — that we should go right up to the highest peak of the Island when we were at it — but thay all tired out before thay got near that length. Mrs R says she has only a bit of a lung, that is

> the reason she is always roaming about Queensland, as she cant stand the climate of Melbourne — it must be a splendid bit — as she was the only one who got to the top of the hill with a native and me — while there she took two outline sketches — I had really at times to hold her or she would have been blown right away — the wind was so strong. [80]

Ellis enjoyed her stay on the lush and fertile island. She tried the native food — dugong, yams, coconut milk and, best of all, pawpaw, as Bruce observed:

> Sanna told me that wherever she goes she always wants Paw Paw apples. I believe she washes her face with the inside of the skin and that accounts for the peach like skin. [81]

Ellis claimed that she put on "a whole stone in weight" during her three-week stay at Murray Island. [82] She also had a productive time painting (cat. no. 95), hampered only by the heat — "for you live in a perpetual state of Turkish bath" [83] — and by the crowds of natives who soon gathered around her and Tom Roberts whenever they began

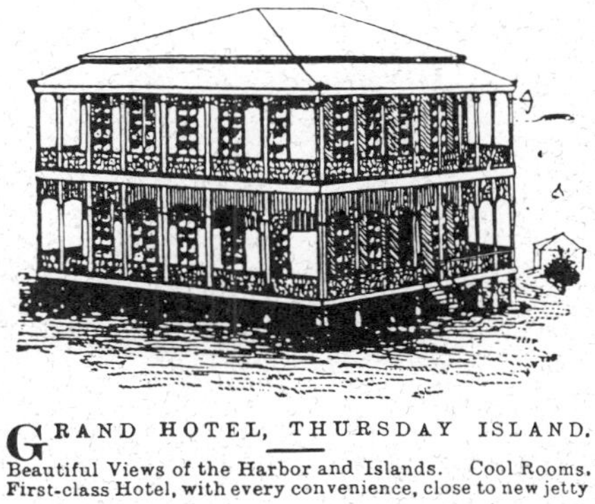

(*Australian Town and Country Journal*, 31 December 1892)

to paint. Tom Roberts was painting on the beach when, Ellis reported (without a tear):

> Just as the forty-guinea finishing touches were being put on the picture, the whole thing fell, butter side downwards, into the sand. [84]

No doubt she had mixed feelings about a fellow competitor who had protested against her award of a gold medal at the Melbourne's Centennial International Exhibition. Roberts, in turn, did not consider Ellis's company worth mentioning when he wrote at length of his visit to the north. [85]

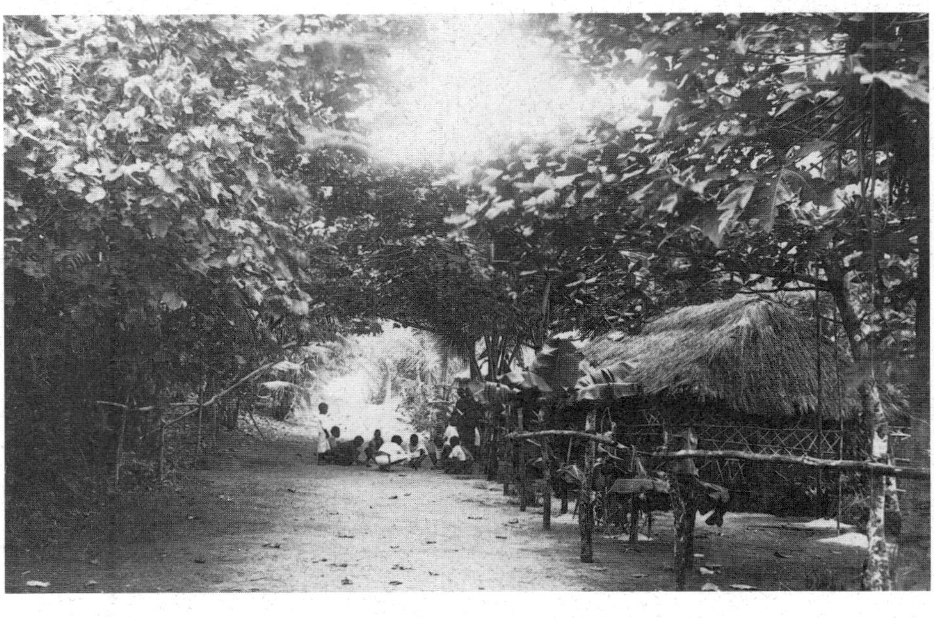

Murray Island, where native food — bananas, coconuts, pawpaws and yams — was "in abundance everywhere", 1899. (John Oxley Library, 49788)

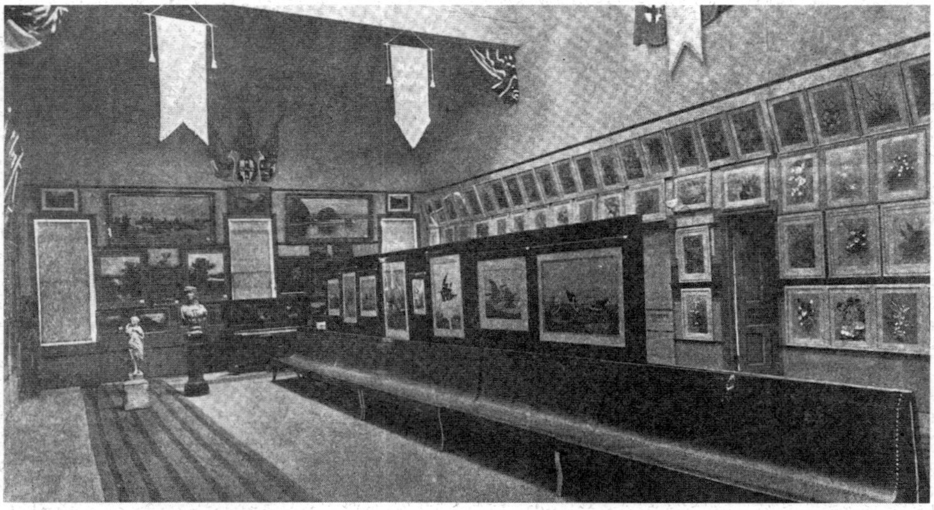

The hall of "Australia House" at the World's Columbian Exposition, Chicago, in 1893. Ellis Rowan's flower paintings are clearly visible on the wall to the right.
(*New South Wales Votes and Proceedings of the Legislative Assembly, 1894-95*)

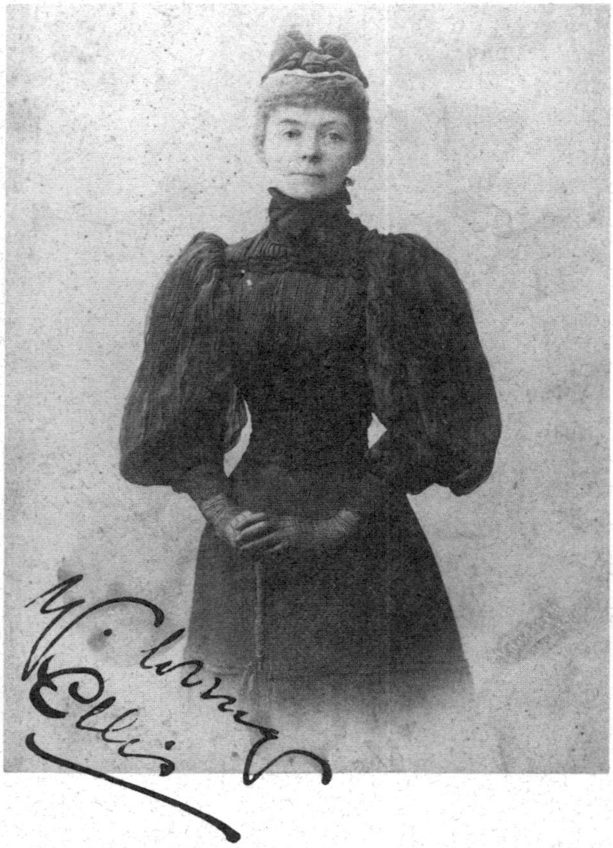

Ellis Rowan the young widow. (Courtesy of the artist's family)

From Murray Island, the *Albatross* and party cruised on to Darnley, Stephen's, and York Islands,[86] arriving at Thursday Island on 3 September.[87] Ellis returned to the Grand Hotel for a final fortnight of fun "what with two dances, races, and a theatre going on",[88] and she must have looked a picture in "pink with black lace" at the Thursday Island Races Ball.[89] On 20 September she arrived back at Cooktown on the same ship as Tom Roberts.[90] Ellis later broke her southward journey at Townsville.[91] She left there on 10 October on the *Wodonga*,[92] stopping off at Sydney to secure her place in the New South Wales Government's contribution to the forthcoming international exhibition at Chicago in America.[93] She was back in Melbourne only a short while before the untimely death of her husband.

At the World's Columbian Exposition at Chicago in 1893 Ellis showed 99 of her watercolour paintings in the fine display of art which hung in the hall of "Australia House", the New South Wales building at the exhibition. Thirty-three paintings were of Queensland subjects[94] but, due to a downturn in the colony's finances, Queensland was not officially represented at the Exposition. A reviewer commented that her paintings, all of which were available for sale, were "a source of attraction to visitors from all parts of the world".[95]

POLITICKING IN QUEENSLAND

1911

Ellis was 63 when she began her second series of visits to Queensland. By then her name as a flower painter was "almost a household word" [1] in Australia and she was determined to win official recognition for her lifetime's work.

She passed through Brisbane early in June 1911, issuing a press release to announce her purpose "to penetrate deeply into the tangled scrubs of the north in search of the flowers she loves". [2] The press release, which found its way into Rockhampton papers on her journey northwards, [3] also announced her wish to keep her collection of paintings in her beloved Australia despite the earlier tempting offer from Germany. "Mrs Rowan's work has been recognised", the press release modestly continued, "by scientific bodies in Britain, Europe and America, as among the best of any living flower artist".

On about 16 June Ellis arrived at Rockhampton by train and, after a few days' stopover, caught another train inland to Longreach. [4] She had travelled to Townsville by 10 July when she departed northwards on the *Kuranda* ,[5] the same steamer which later brought her back to Townsville on 6 October. [6] It may have been during this time that she visited F. P. Dodd's famous collection of butterflies and moths at Kuranda, a local tourist attraction after his arrival there in 1904, for Ellis was said to have copied insects from his collection. [7] She certainly visited Cardwell, [8] where she stayed some time at the Marine Hotel when it was a simple single-storied building, not quite to her taste. The proprietress Mrs Maria Cook was amused that Ellis wanted to leave immediately — there was a week's wait until the next ship and Ellis found herself a captive guest.[9]

At the Murray River north of Cardwell Ellis painted a large spotted orchid, "Mrs Rowan's Phaius", named *"Phaius grandifolius* var. *Rowanice"* by Colonial Botanist F. Manson Bailey (1827-1915) [10] in her honour, [11] but subsequently found to have been described already by Ferdinand von Mueller and named *Phaius australis*. Ellis also found a "most picturesque garden" of specimens to paint at Lucinda Point [12] near Dungeness (cat. no. 31), and visited the well-known author and naturalist E. J. Banfield [13] and his wife Bertha at their "perfect paradise" on Dunk Island. [14]

In November Ellis passed through Rockhampton on her way back from her extensive visit to the north and west. She encountered — or rather, accosted — the Premier Digby Denham (1859-1944) [15] somewhere between Gympie and Brisbane, where he intervened to obtain her some spectacular yellow *Banksia* which she had seen growing beside the railway line (possibly cat. nos 48, 53 or 69). Consequently, huge clumps, together with "an immense quantity of soil and other impedimenta", kept arriving day after day at her hotel — "and the artist incontinently fled". [16]

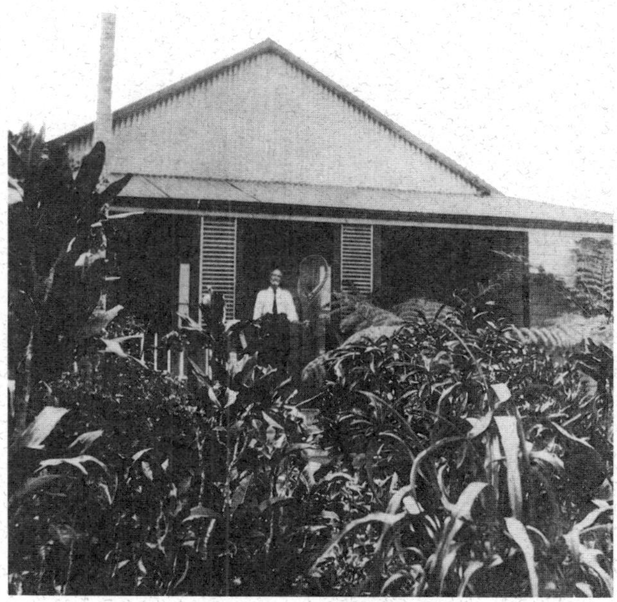

F. P. Dodd on the verandah of his cottage at Kuranda, c.1920s. (Queensland Museum, courtesy of the Dodd family)

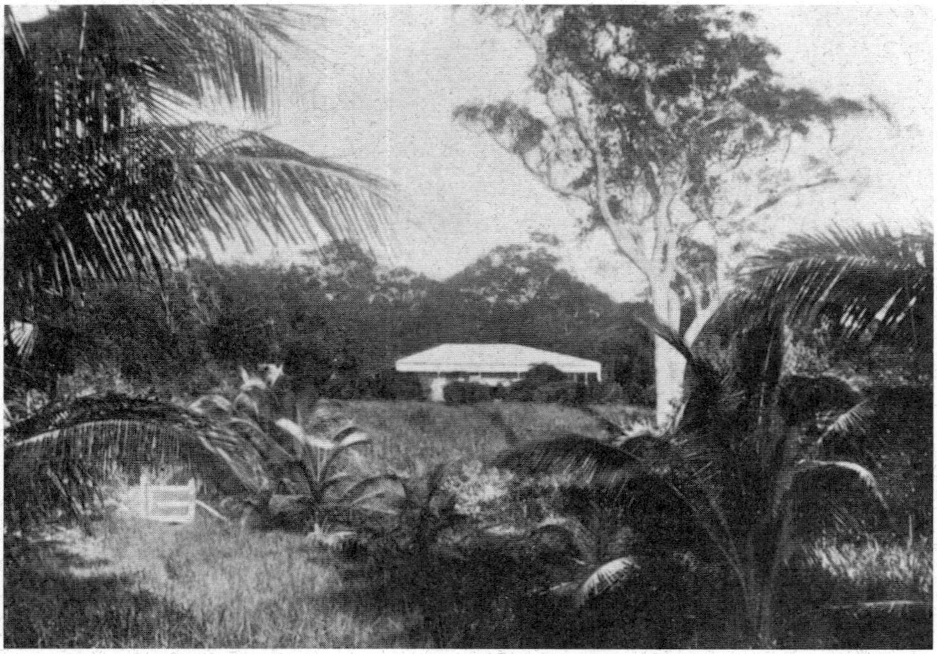

The Banfield bungalow at Dunk Island.
(E. J. Banfield, *My Tropic Isle*, T. Fisher Unwin, London, 1911.)

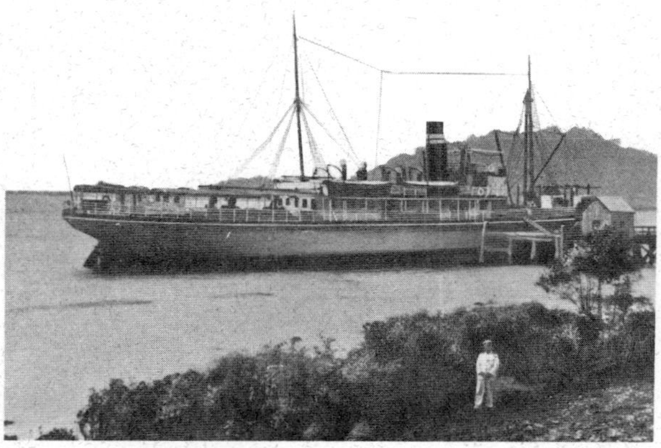

Above: *The AUSN Company's <u>Kuranda</u> in Mourilyan Harbour. Built in 1901, the Kuranda was a favourite in the northern outposts trade.* (John Oxley Library, *30082*)

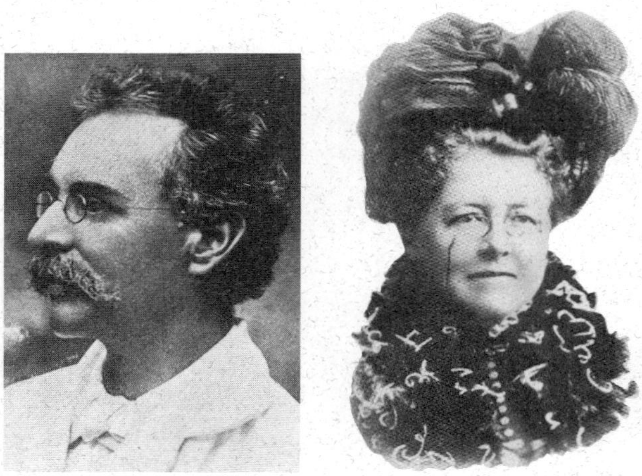

E. J. Banfield the self-styled "beachcomber" and his wife Bertha, 1901. (E. J. Banfield, *Last Leaves from Dunk Island*, Angus and Robertson, Sydney, 1925.)

No doubt the Premier also was asked to facilitate her proposed flower-hunting expeditions in the south, around the Blackall and Macpherson Ranges, because he wrote to enlist Bailey's advice on the best localities for flowers. [17]

During her 1911 visit Ellis established a useful rapport with the kindly old botanist and sought his approval of her paintings. Bailey's approval was qualified — he would have preferred the paintings to also show the anatomy of the plants,[18] as was standard in traditional botanical illustration. However, he regarded the paintings as accurate enough to choose sixteen from her 1911 visit for colour plates [19] in the massive work of his later career, the *Comprehensive Catalogue of Queensland Plants,* published in 1913. Ellis's colour plates make an artistic addition to the book's 970 strictly didactic line figures, mostly the work of Bailey's grandson Cyril Tenison White (1890-1950) [20] who had joined him as a pupil-assistant in 1905. By 1911 Bailey was near the end of his long and productive career as Colonial Botanist and his grandson, later to succeed him as Government Botanist, was assisting him with clerical duties and undertaking fieldwork.

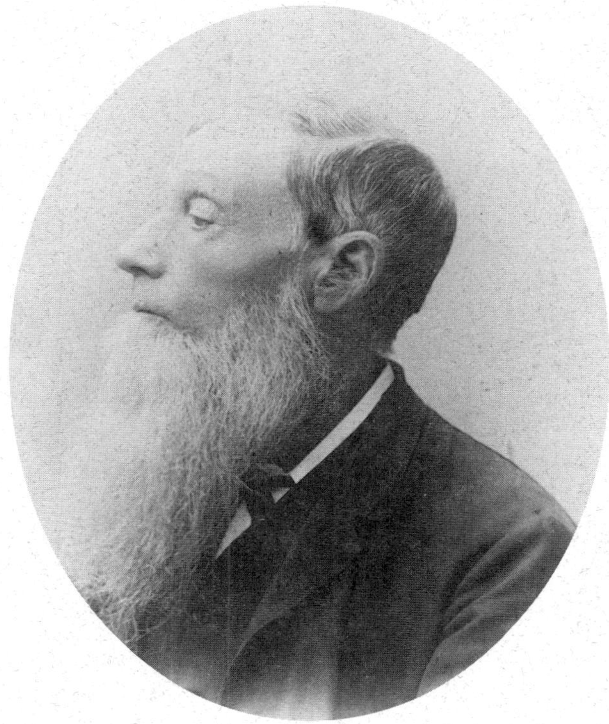

Frederick Manson Bailey, CMG, FLS, the much esteemed Colonial Botanist of Queensland. Bailey always wished to retain his original title of "Colonial", in preference to "Government" Botanist. (John Oxley Library, 160401)

It may have been in response to Ellis's request for guidance that C. T. White, in company with the Government Entomologist Henry Tryon and local resident Romeo Lahey, made a field trip to the higher parts of the Macpherson Range at about the end of 1911. Several of the plants collected and described on that trip [21] were subjects of Ellis's paintings from 1911, [22] but it has not been possible to prove that she joined the field trip. Bailey might also have directed her to the rich rainforests from Eudlo to Cooroy, which had been a favourite collecting ground of his in the early 1890s.

On 6 December Ellis wrote to the Premier inviting him to inspect her paintings which were then on view at Bailey's office in the Agriculture Building and, to prove that she was no mere painter of *Banksia,* she enclosed recommendations "from the principal Botanists of the World." [23] Denham was then preoccupied with a major constitutional crisis which threatened to split the governing Liberal Party and did not have any time to spare for Ellis. By 23 December, on the eve of her departure from Queensland she had lost patience and wrote again to the Premier:

> I am not easily beaten and perhaps later on you may have more time to go into the matter. I have promised to return in three months and hold an exhibition and in this way test the wishes of the general public — Please forgive me for adding to your troubles at such a time. I did not mean it. You are an ambitious man (and rightly so) and I as a woman am just the same and some day I will look forward to the time when my own country will recognise that I have not worked for it in vain. [24]

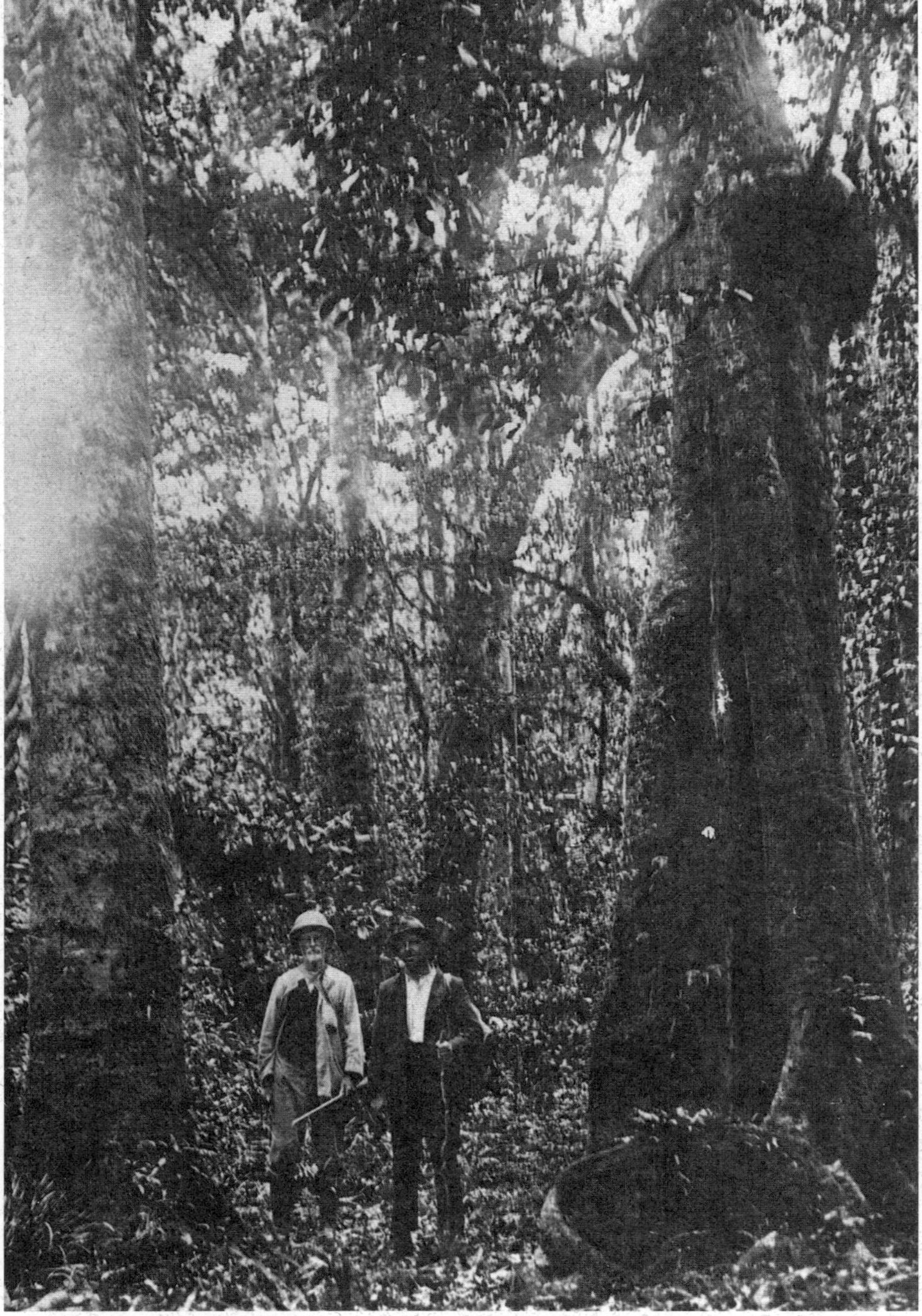

Henry Tryon (left) *and C. T. White* (right) *at the Macpherson Range, 1918.* (Queensland Herbarium Archives)

1912

As promised, Ellis returned to Brisbane in the following year to stage the exhibition of her Queensland paintings. She arrived early in July and stayed at the Belle Vue Hotel together with her old friend Mrs Evelyn Casey.[25] The Caseys were then fellow residents of Melbourne, but frequently visited Queensland following R. G. Casey Senior's appointment in 1899 as a director of the Mount Morgan Gold Mining Company. Mrs Casey later went to visit her sister, the popular society hostess Mrs Edith Lumley Hill,[26] at Bellevue Station in the Brisbane Valley[27] before returning to Melbourne, while Ellis remained in Brisbane to prepare her exhibition.

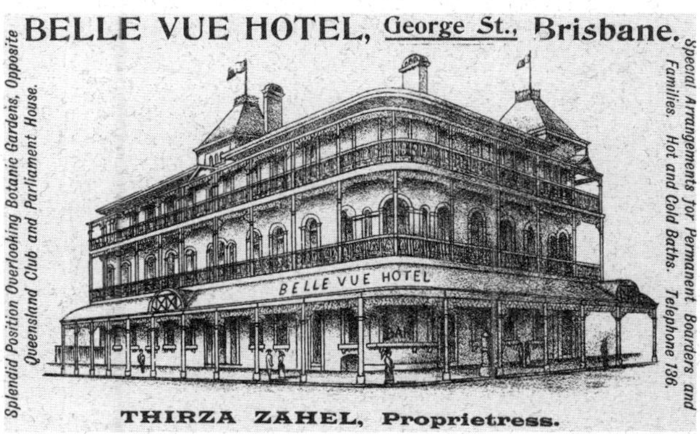

Ellis stayed here before and during her exhibition of 1912. (*Queensland Illustrated*, 1895-96)

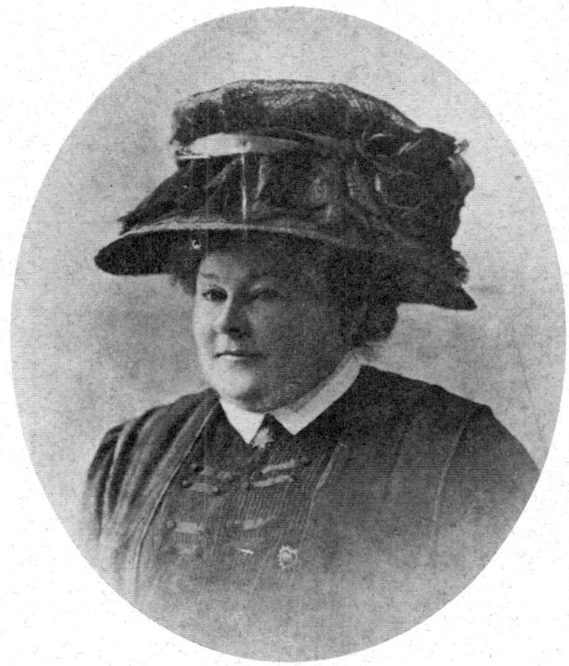

Mrs Edith Lumley Hill of Bellevue Station, whom Ellis described as "a most exceptional woman and added to her great charm she is so clever". (John Oxley Library, 157309)

She sought the assistance of Colonial Botanist Bailey in naming her paintings. He named all but several that could not be identified without plant specimens.[28] Though Bailey provided the identifications, the writing of his assistant C. T. White[29] appears on the reverse of many of the paintings that had not already been named by von Mueller (see Catalogue, pp. 65-81). The unframed paintings were framed by A. L. Hambleton, of Brisbane, whose label was fixed to many of the original frames. Ellis began her campaign to make the State Government purchase her Queensland collection[30] with another reminder of her "five figures" offer from Germany, and announced that "it would appear that no pressure should be necessary" to secure the sale.[31] Meanwhile, she was stirring up as much public pressure as possible.

The exhibition, hung in a large upstairs room in Brisbane's old Town Hall, opened on Monday 5 August with an "At home" when Ellis the matron of honour, sporting a large white ostrich plume in her hat, entertained her invited guests — parliamentarians (for she admitted that the event was "really only for them"[32]) and leading citizens.[33]

The old Brisbane Town Hall in Queen Street, at about the time of Ellis Rowan's exhibition. (John Oxley Library, 21993)

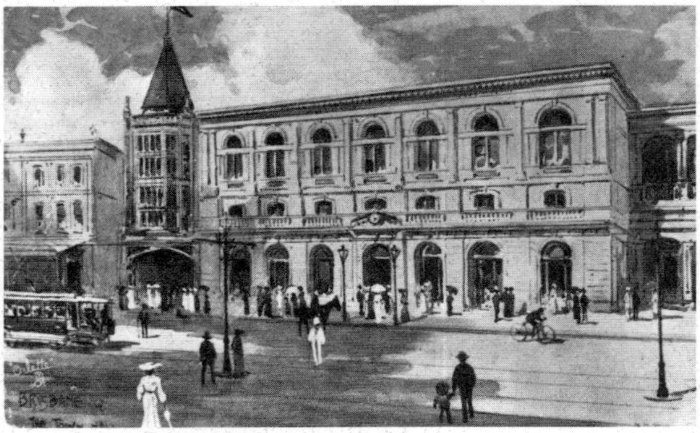

The exhibition remained open daily for over a month and was timed to coincide with the National Association of Queensland's annual show which brought the country to town. Among the visitors to town was Mrs Lumley Hill who was sent by Ellis to meet the Premier's Under Secretary [34] — no doubt to discuss the merits of Ellis's paintings rather than of Bellevue Station's prize-winning Hereford herd.

Back at her exhibition, the artist sat at a little table all day long, collecting the (considerable) one shilling admission charges and answering visitors' questions. One visitor remarked:

> She is a study in herself which makes one stop and think, realising that in the character of this "lady of the flowers" there is not only a great artist but a very great woman. [35]

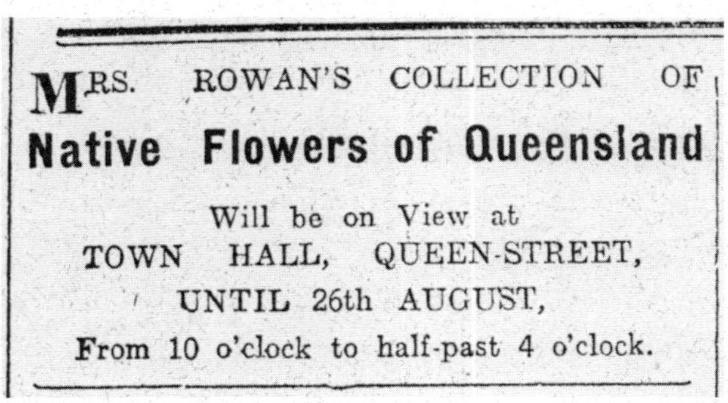

MRS. ROWAN'S COLLECTION OF

Native Flowers of Queensland

Will be on View at

TOWN HALL, QUEEN-STREET,

UNTIL 26th AUGUST,

From 10 o'clock to half-past 4 o'clock.

(*Daily Mail* (Brisbane), 17 August 1912)

Notices of the exhibition appeared with monotonous regularity in the social columns of the daily papers, including Ellis's paid advertisements: "Mrs Rowan's Exhibition of Native Flora of Queensland is still attracting much attention"; [36] and "Mrs Ellis Rowan's collection of paintings . . . attracted a large number of visitors yesterday, including . . ." [37] — notables being added accordingly. The exhibition was well attended by the general public, but a fortnight passed before Government Ministers came to inspect the collection on offer, [38] though the Premier was ill during that time.

What finally stirred them into action was a letter written by A. B. Wilson, [39] President of the Queensland Institute of Architects. The letter, published in the *Brisbane Courier* of 22 August, [40] stressed the educational value of the collection. It would provide, Wilson argued, an "invaluable stimulus" to local students of design and so encourage a uniquely Australian style of decorative design — then a vital concern of many art schools and technical colleges. Within a week of Wilson's influential letter most of the Government Ministers had "visited and admired" the collection. [41] Then other interested parties made public appeals to the Government to purchase the collection: the Queensland Naturalists' Club; [42] the Horticultural Society of Queensland; [43] Henry Tryon[44] and Sydney B. J. Skertchly, [45] both leading lights in the Naturalists' Club; and various other patriots who vouched that paintings of the State's magnificent flora would uplift the souls of all true Queenslanders. A local music teacher-cum-composer, R. Randolph Arndell of New Farm, was moved to write a gavotte, *"Chez Madame"*, in honour of Ellis. [46]

The Premier Digby Denham eventually gave way — he had forcibly suppressed Brisbane's rioting workers during a general strike early in 1912, but he was unable to withstand the pressure of Ellis and her Establishment supporters. On 14 September she was officially notified of the Government's decision to purchase her Queensland

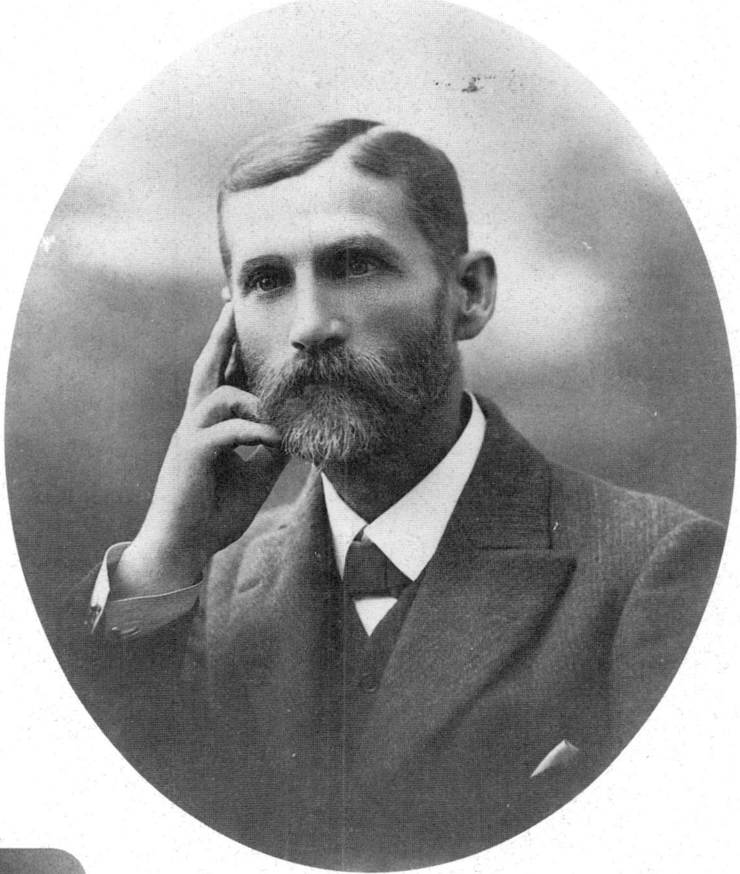

Digby Denham, the Premier who put down the general strike of 1912 and acquired the Ellis Rowan Collection. (John Oxley Library, 68206)

Alexander Brown Wilson, President of the Queensland Institute of Architects for many years from 1899 to 1922 and influential advocate for the Ellis Rowan Collection. (Courtesy of Mr Blair Wilson)

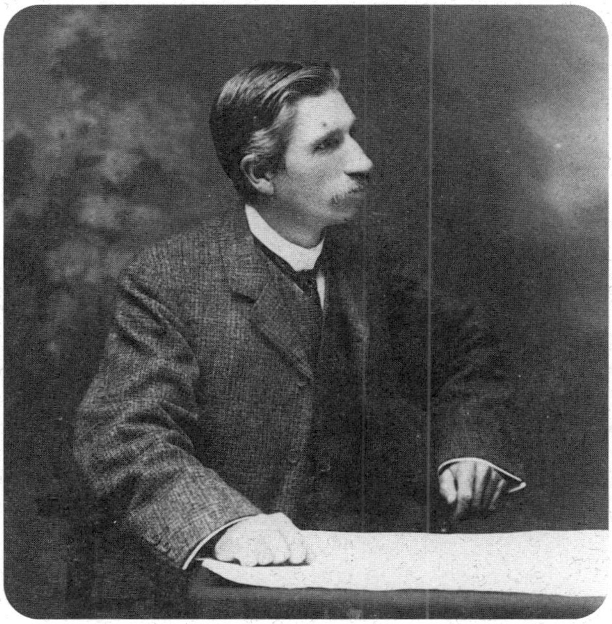

collection for £1050.[47] She issued reports that the Government was purchasing 100 of her paintings but that she, "with laudable patriotism", was presenting the remaining 25 so that the "full collection" would stay in Queensland.[48] Her patriotism was rather questionable given that "many" of the exhibition's 154 paintings had already been sold at the opening event alone,[49] not to mention the countless other Queensland paintings sold from earlier exhibitions (see Appendix 1, pp. 84–85). Moreover, the purchase price represented a generous 81 per cent of the Premier's annual salary for 1912.

Ellis and her collection remained in the spotlight for months after her victory was secured, while a debate raged in the Brisbane press over the fate of her paintings: were they works of "art" worthy of the State's Art Gallery or were they botanical specimens more appropriate to the Museum? The main protagonists in the debate were art connoisseur Dr Harry M. Lightoller, who pronounced Ellis's paintings "quite unsuited for an art gallery"[50] and Ellis's friend, the Sydney botanist and educationalist, Sarah Hynes, who sprang to her defence:

> ... That artist ... can quite afford to ignore such an opinion seeing that her work has brought her fame and honour in all parts of the world. Mrs Rowan has won medals and certificates enough to make any woman vain, but she is not built that way ...[51]

Miss Hynes, B.A., had the last word in the continuing debate and silenced the good doctor, himself a keen amateur painter:

> ... This question has been raised altogether too late ... Mrs Rowan's pictures hold an honoured place in the equally first-class, if not more so, art galleries of Europe, Africa, America, and also of the sister States of Australia. Why should Brisbane, then, be an exception? In conclusion I may state that if there were less ego in H. M. Lightoller's cosmos he would know that the opinion of a mere amateur carries very little weight.[52]

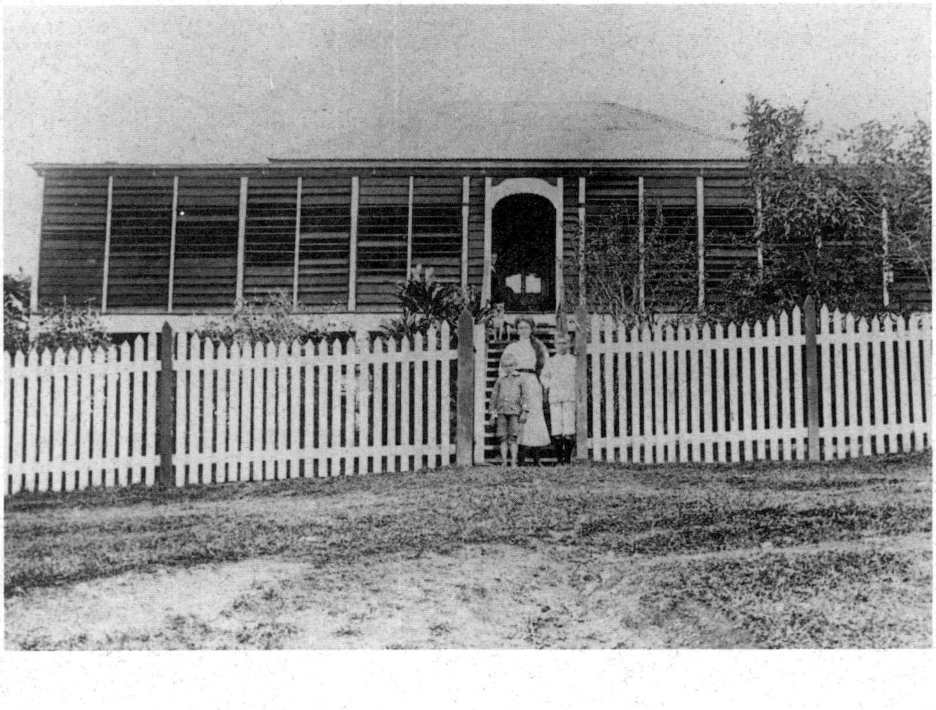

The MacArthur residence in Rankin Street, Innisfail, where Ellis Rowan stayed in c.1912. The MacArthur children are in the garden. (Courtesy of Mr Peter Croker)

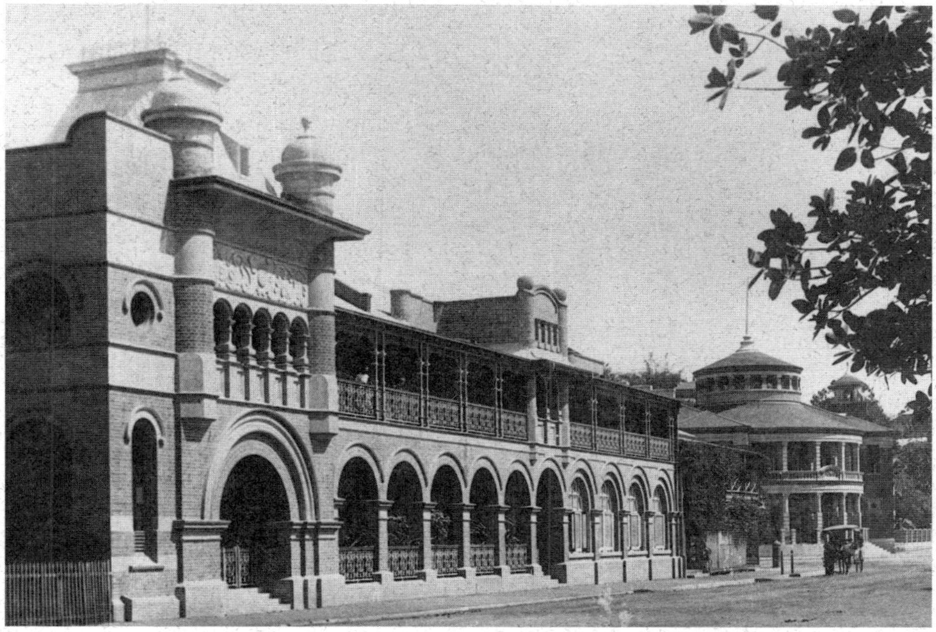

Queen's Hotel on The Strand, Townsville — "the rendezvous of the north" — where Ellis stayed in 1912. (John Oxley Library, 43607)

The Government's decision to house the collection in the Queensland Museum was mainly due to the Art Gallery's lack of space to display even its existing works,[53] for the Gallery was then located in a crowded top-storey room of the Land Administration Building. The move to the Museum brought many public protests. One indignant correspondent wrote that "real works of art" should not be placed with "musty" Museum artifacts.[54] Another correspondent complained that the collection would be far less accessible at the Museum, then housed in the old Exhibition Building at Bowen Park, than in the city.[55] The Government's decision in favour of the Museum also took into account the advice of the Art Gallery's Honorary Curator, Godfrey Rivers,[56] who saw the paintings as botanical specimens and quite unfit for his albeit crowded walls.[57] His opinion was predictable in view of Ellis's aloofness from the local art community, just as she had distanced herself from the professional artists of Melbourne. Moreover, genteel flower painting was hardly a serious pursuit for Rivers, a professional art master trained at the Slade School of Fine Art in London. The Queensland Art Society, of which Rivers was a leading member and past president, remained conspicuously silent during the campaign for the purchase of Ellis's collection.

While the fate of her collection was being so hotly debated Ellis was busily flower-hunting in the north, as she was yet to exhaust Queensland's vast store of wild flowers to paint. On 21 September, within a week of securing her victory, she had left Brisbane on the *Wyreema* for Cairns,[58] intending to proceed "inland" for a month.[59] While in the north she visited Babinda, from where she despatched three specimens to F. Manson Bailey,[60] and then visited Innisfail[61] and Cardwell.[62] It was probably during this journey that she stayed at Innisfail with Mrs Anniechen MacArthur (née Bauer), her former pupil from Cooktown, and they enjoyed long sessions painting together on the verandah.[63] On 28 November Ellis left Cardwell for Townsville on the *Palmer*,[64] the same ship on which she had first travelled in 1887. At Townsville she again stayed at the Queen's Hotel[65] which

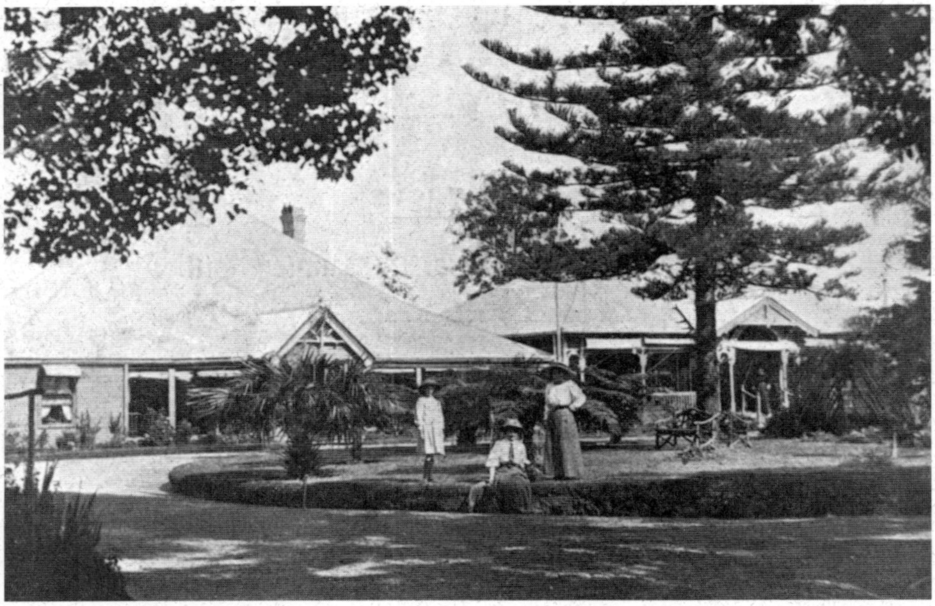

Bellevue homestead with Mrs Lumley Hill's daughters in the foreground. Ellis was one of many guests who enjoyed Mrs Hill's celebrated hospitality. The Prince of Wales, later King Edward VIII, also visited Bellevue. (John Oxley Library, 157312)

had acquired an elegant new building since her previous visits.

Arriving in Brisbane on 5 December on the *Wodonga*,[66] she went to stay with Mrs Lumley Hill at Bellevue Station[67] for a rest — and possibly for a victory celebration — before departing by train for the south.[68] She left declaring her "love" for Queensland and vowing to return in the following year.[69]

1913

As planned, Ellis returned in 1913 for yet more flower-hunting, arriving in Brisbane late in June.[69] She spent some time visiting friends in the south,[70] possibly Edward King and family of Craigend Station at Charlton on the Darling Downs, with whom she is known to have stayed at about that time.[71] She departed for the north on the *Wodonga* on 11 July.[72] Her subsequent movements are uncertain, except that Cooktown was on her itinerary.[73] At Cooktown she stayed with James Gibson and family at Kings Plains Station to paint the magnificent pink waterlilies growing on the lake there.[74] F. Manson Bailey also received specimens of the waterlilies and intended to name them "*Nymphaea Lotus* var. *Gibsonae*", a new variety, after Miss Eleanor Gibson.[75] Later in 1913 Bailey asked Ellis to paint a waterlily especially for one of his regular contributions to the *Queensland Agricultural Journal*,[76] but she was unable to meet his request promptly[77] and Bailey's proposed article did not eventuate. She arrived back in Brisbane on the *Suva* on 14 September,[78] and departed immediately for the south to nurse her sister who was seriously ill.[79]

After this Ellis turned her attention to New Guinea. There is as yet no evidence to suggest that she broke her several journeys further north for more flower-hunting in Queensland.

On the verandah of Mr (?Osmer B.) Ward's home at Samarai, New Guinea, c.1916. Ellis is second from the right. The couple next to her are Reg Fletcher, a local trader and plantation owner, and his new wife Edith. Ellis stayed with the Fletchers at Samarai. (Courtesy of Mrs Joan Karlino)

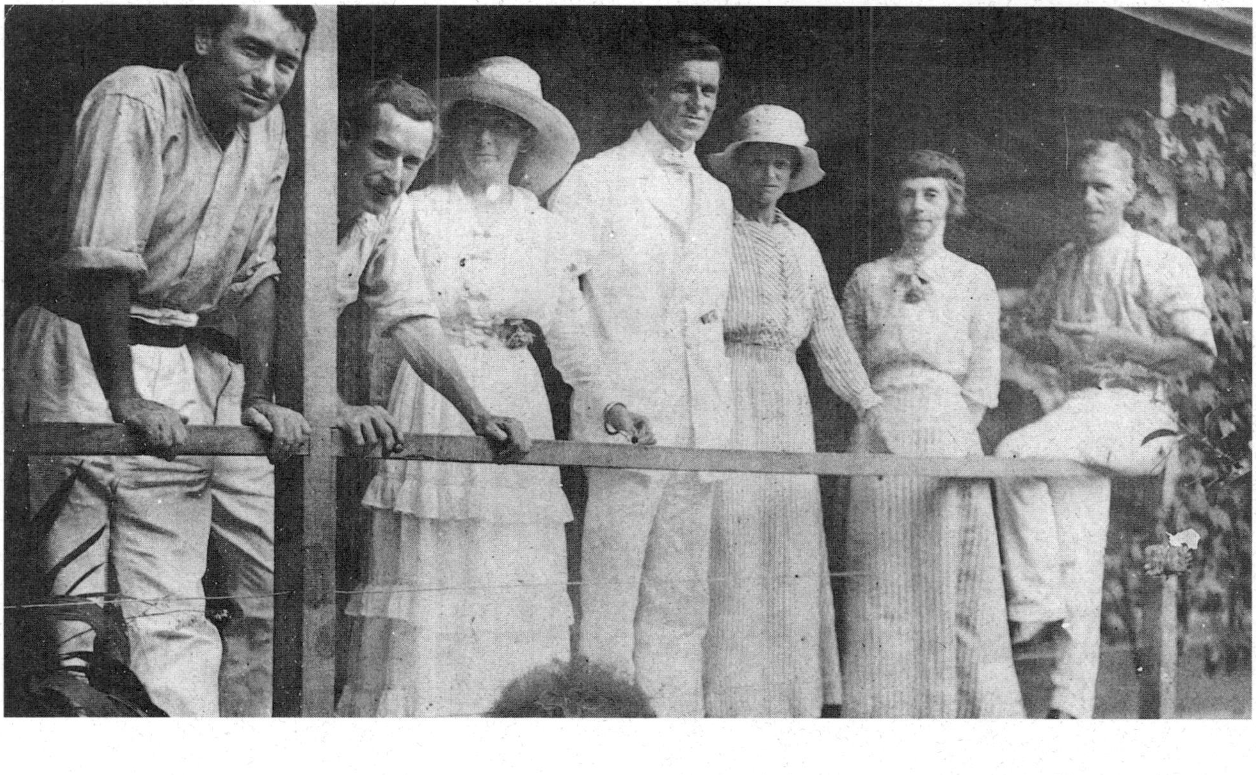

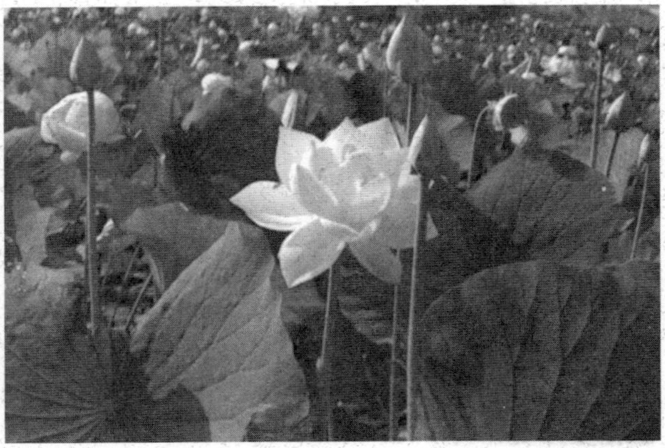

Left: *Waterlilies at Kings Plains Station near Cooktown, a favourite hunting ground of Ellis Rowan.* (Courtesy of Mr Archie Black)

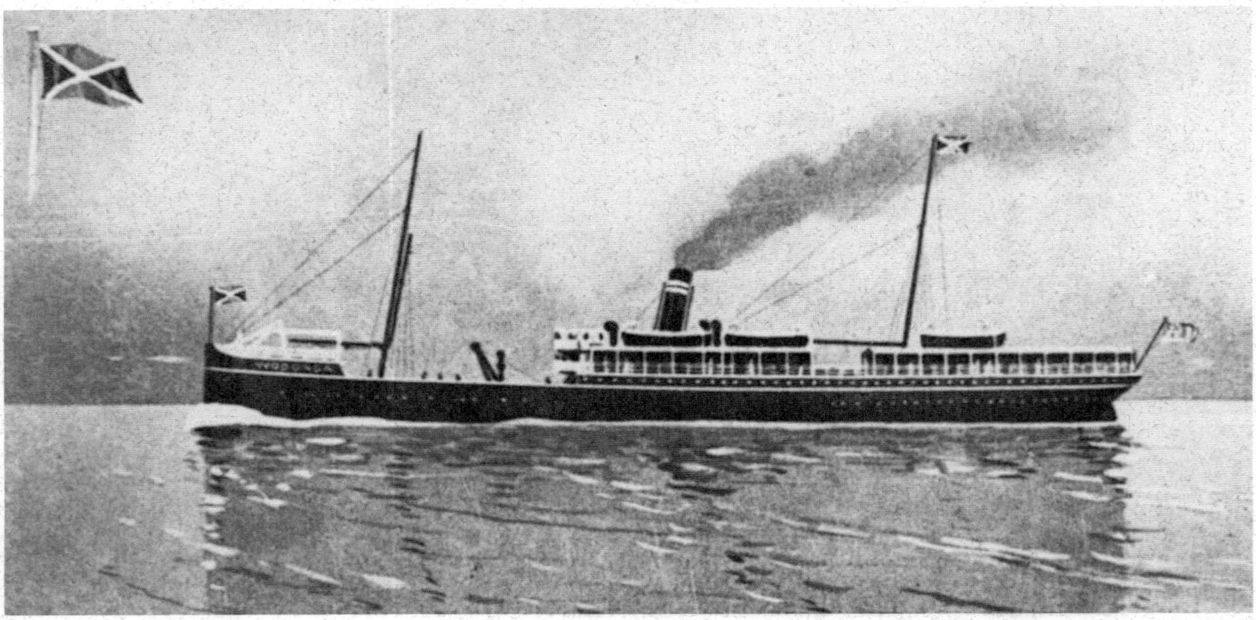

Below: *The Wodonga served the AUSN Company's east coast route from 1890 until until the 1920s.* (John Oxley Library, 61820)

C O L O U R   P L A T E S

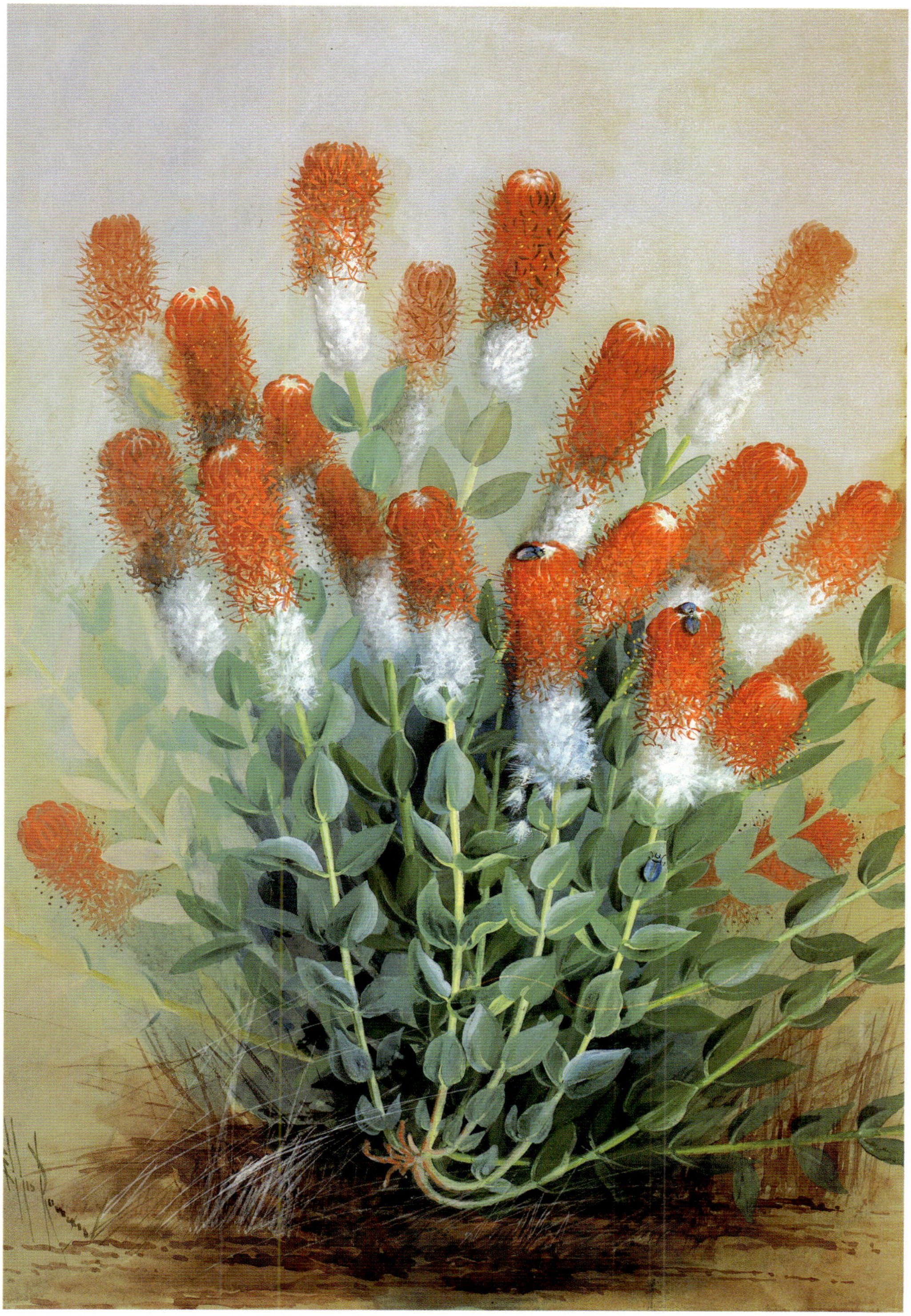

Pimelea decora Domin FLINDERS POPPY (cat. no. 105)

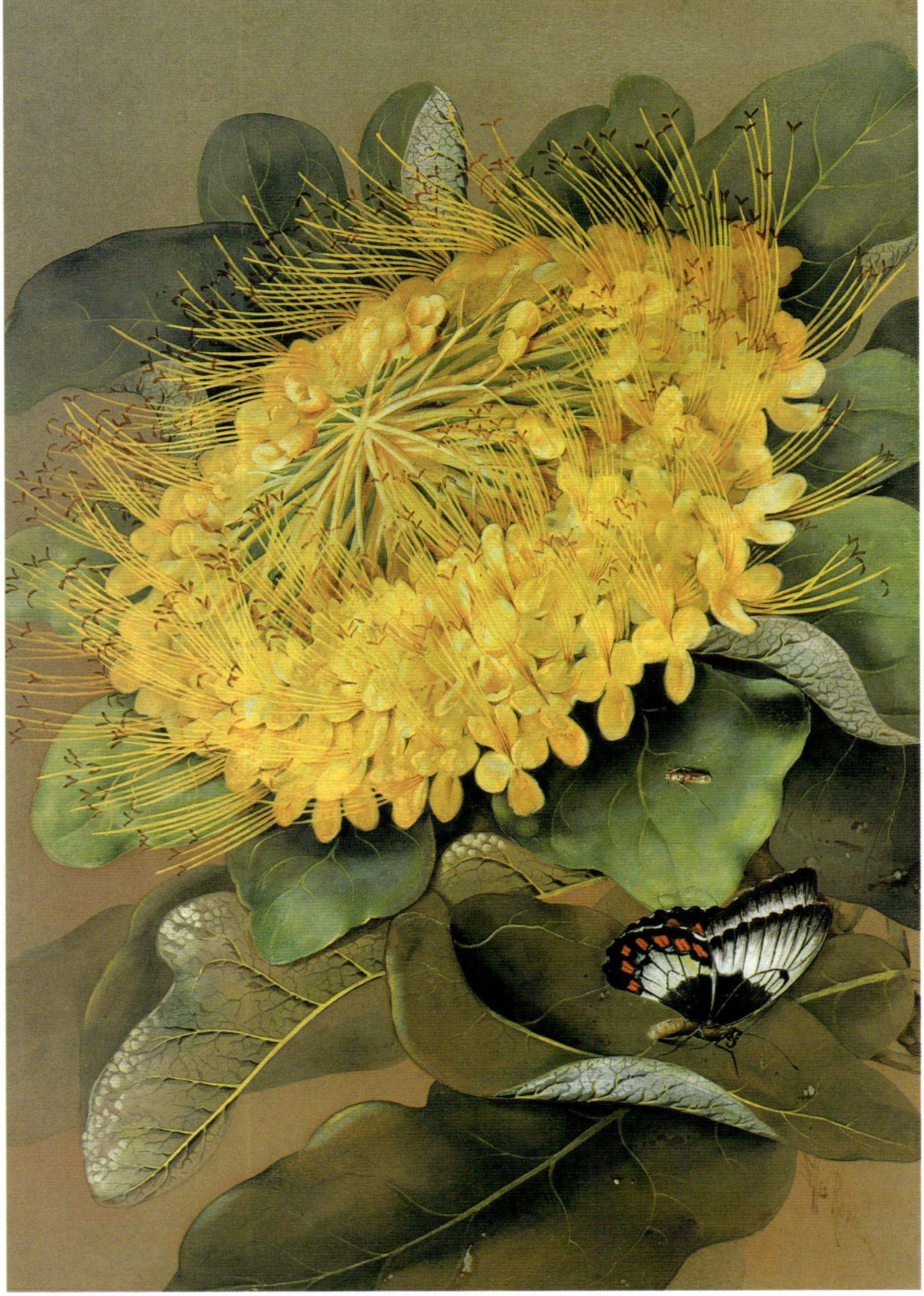

Deplanchea tetraphylla (R. Br.) F. Muell. GOLDEN BOUQUET TREE (cat. no. 63)

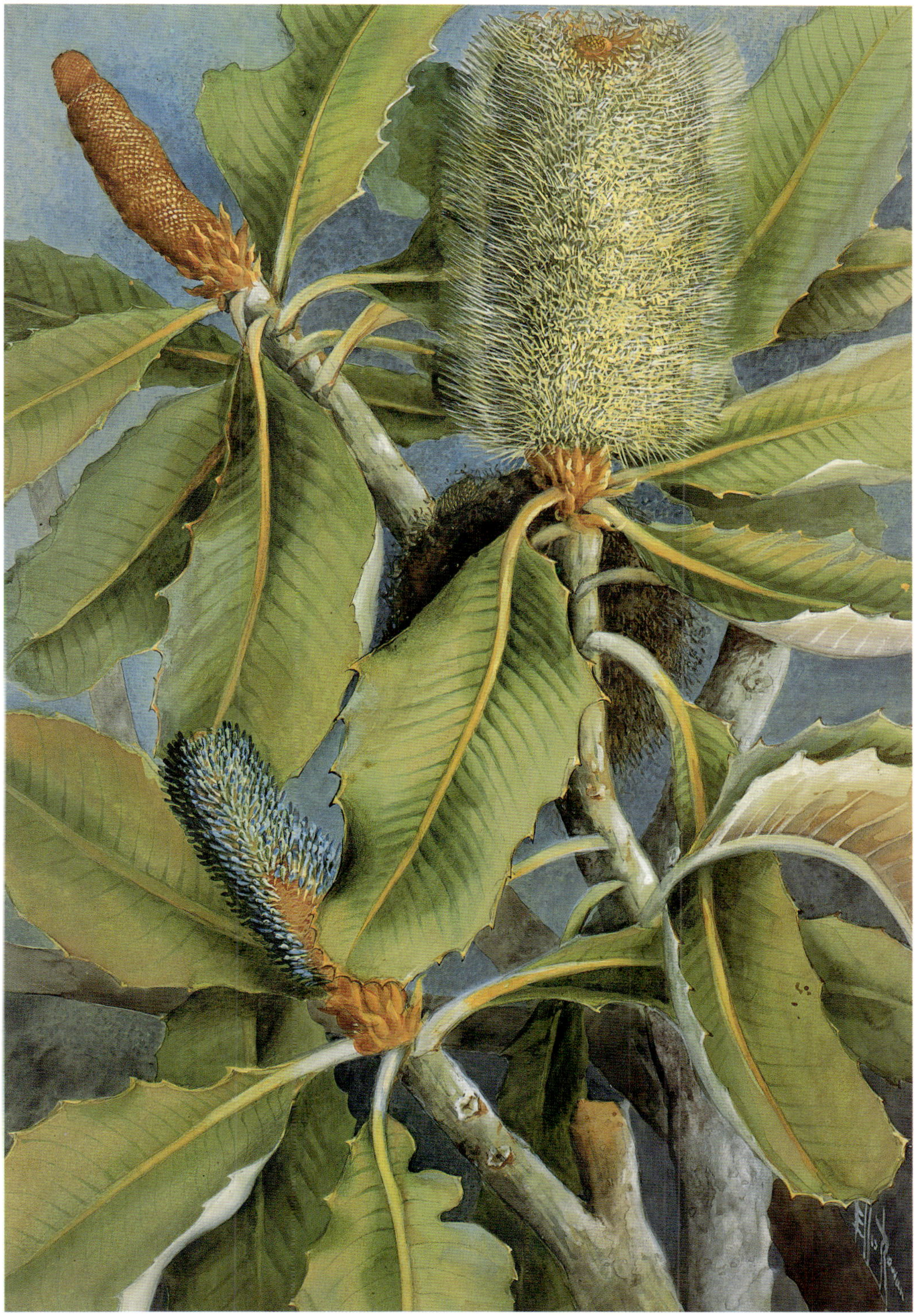

Banksia robur Cav. BROAD-LEAVED BANKSIA (cat. no. 53)

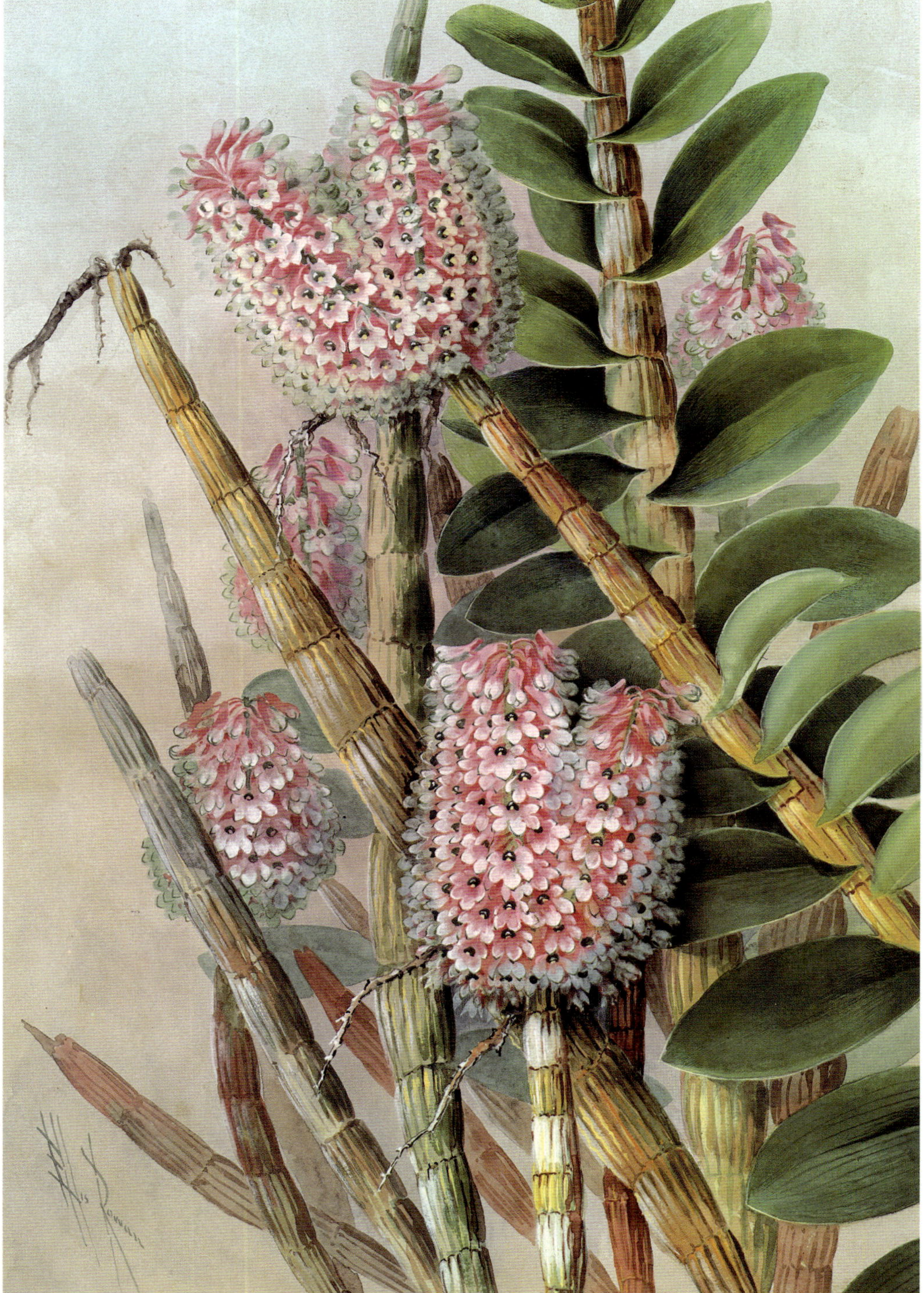

Dendrobium smillieae F. Muell. BOTTLEBRUSH ORCHID (cat. no. 6)

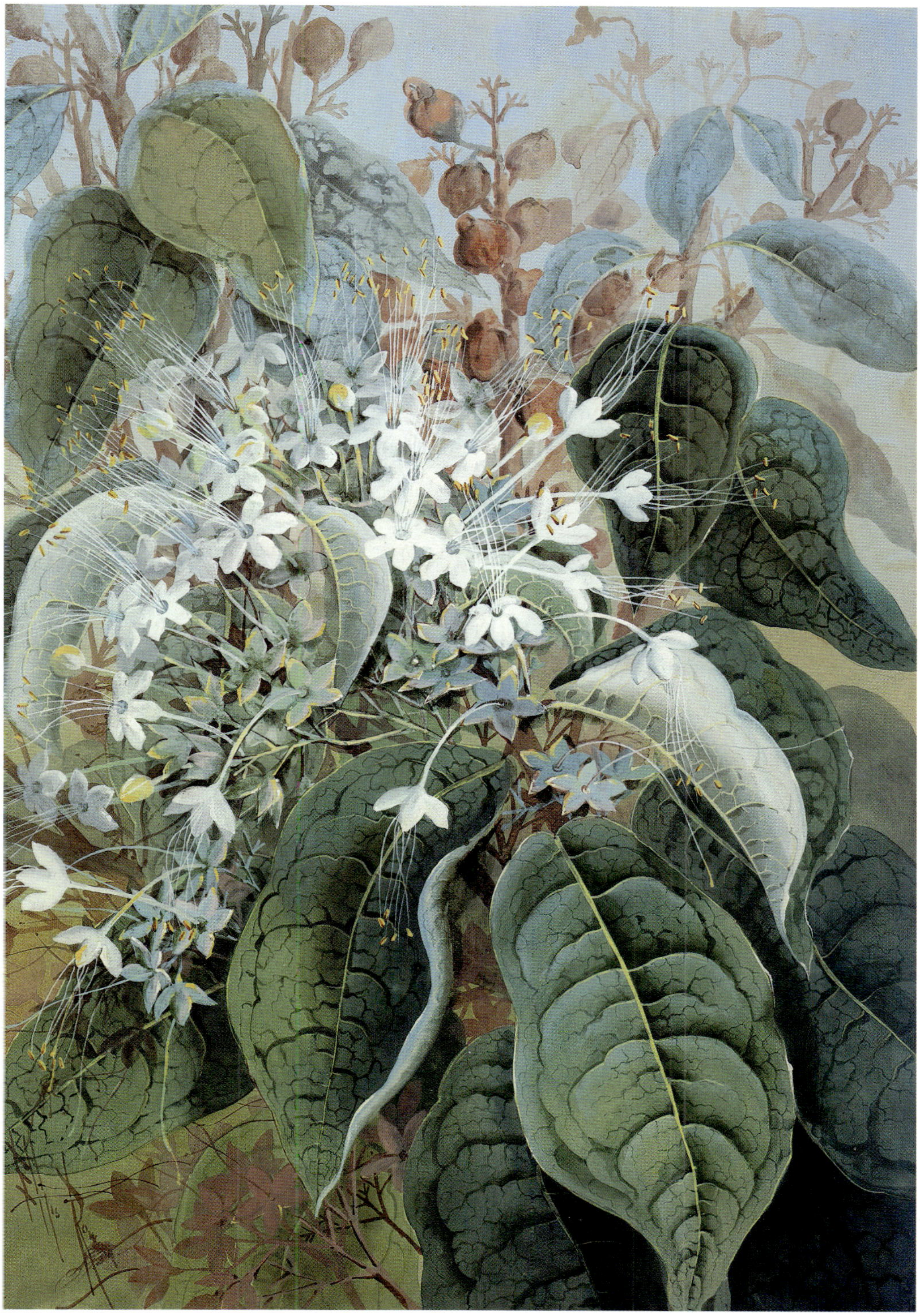

Clerodendrum cunninghamii Benth. NATIVE CLERODENDRUM (cat. no. 36)

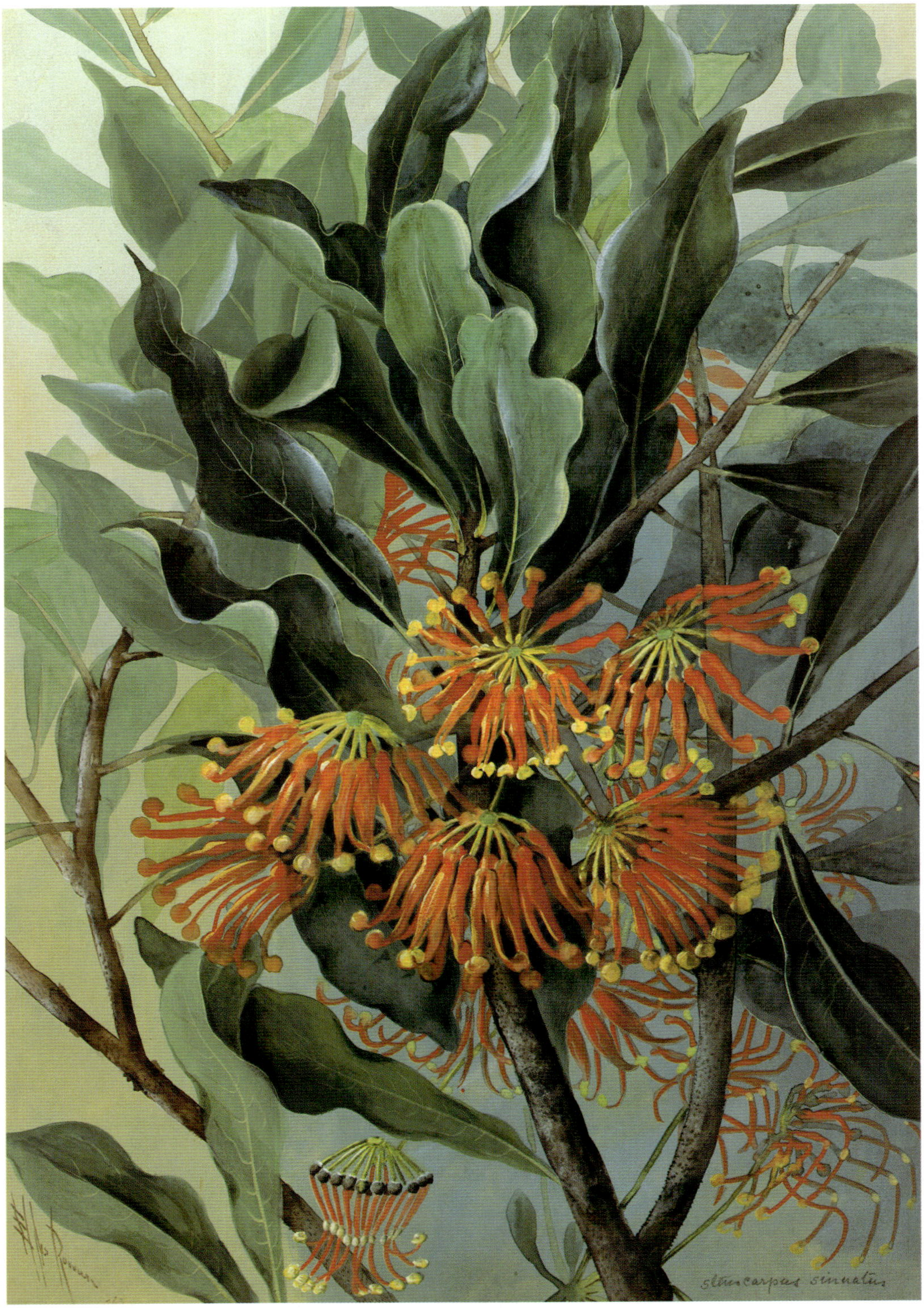

Stenocarpus sinuatus Endl. WHEEL OF FIRE (cat. no. 85)

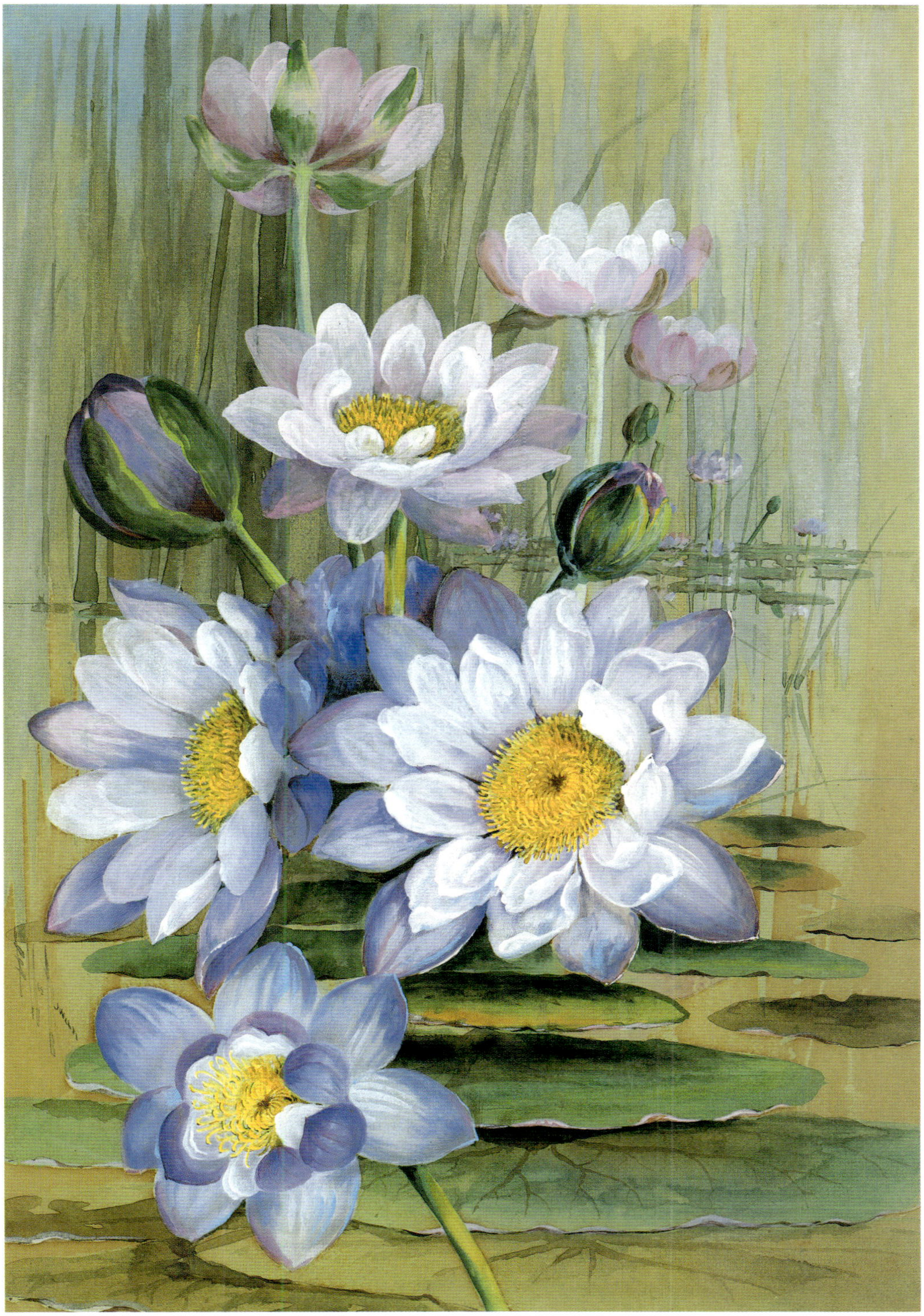

Nymphaea gigantea J. D. Hook. GIANT BLUE WATERLILY (cat. no. 121)

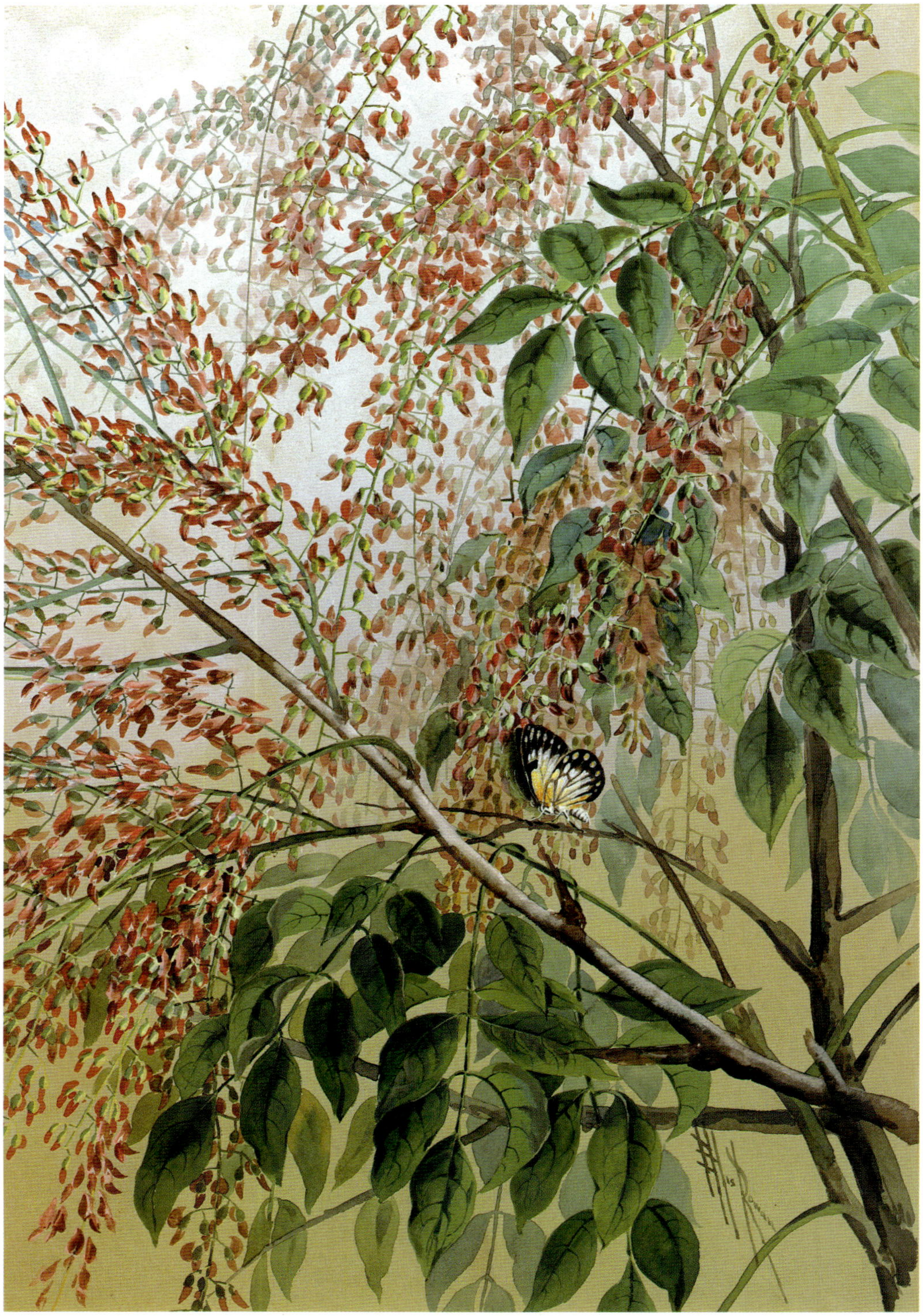

Austrosteenisia blackii (F. Muell.) Gees. BLOOD VINE (cat. no. 20)

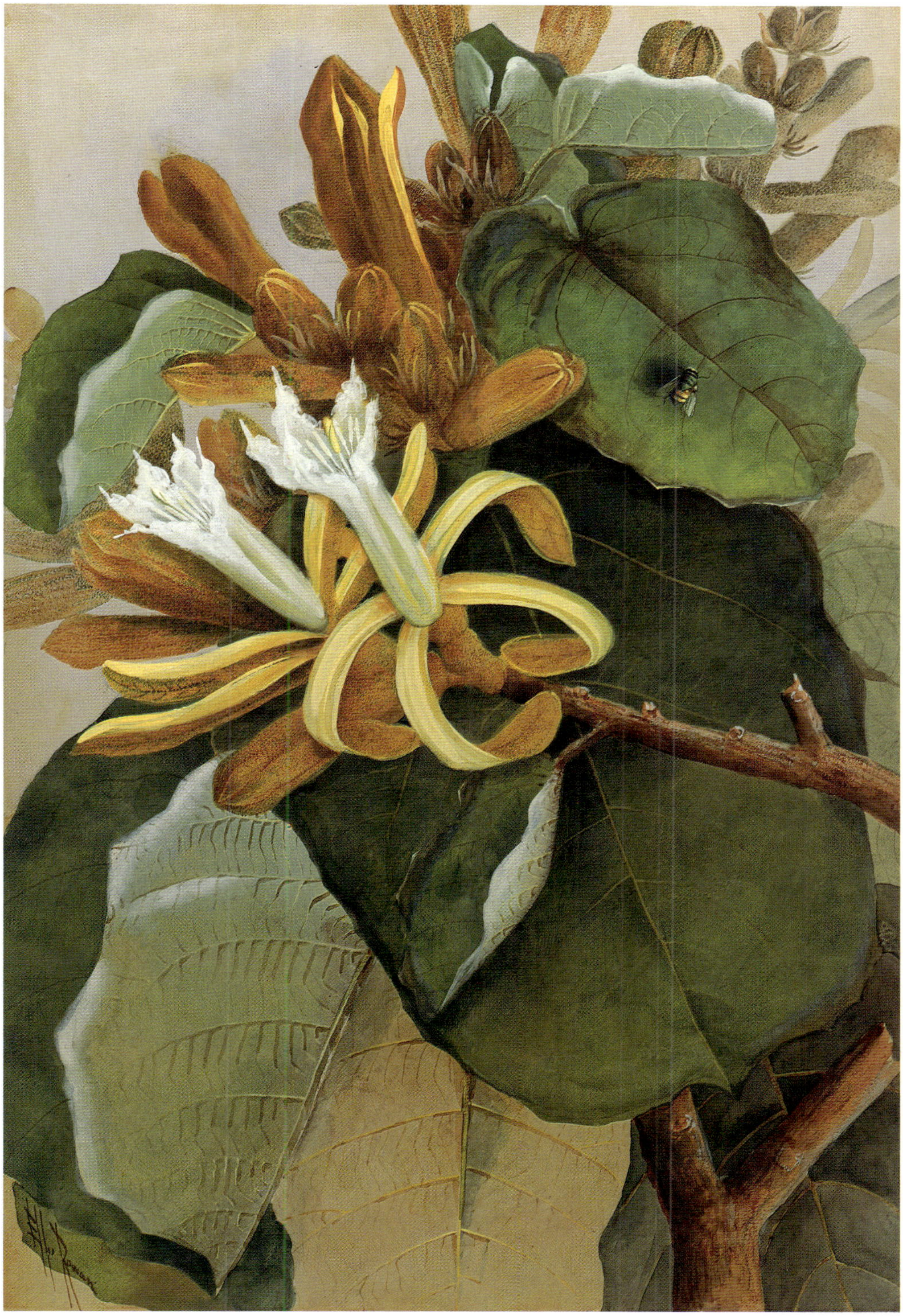

Pterospermum acerifolium Willd. (cat. no. 116)

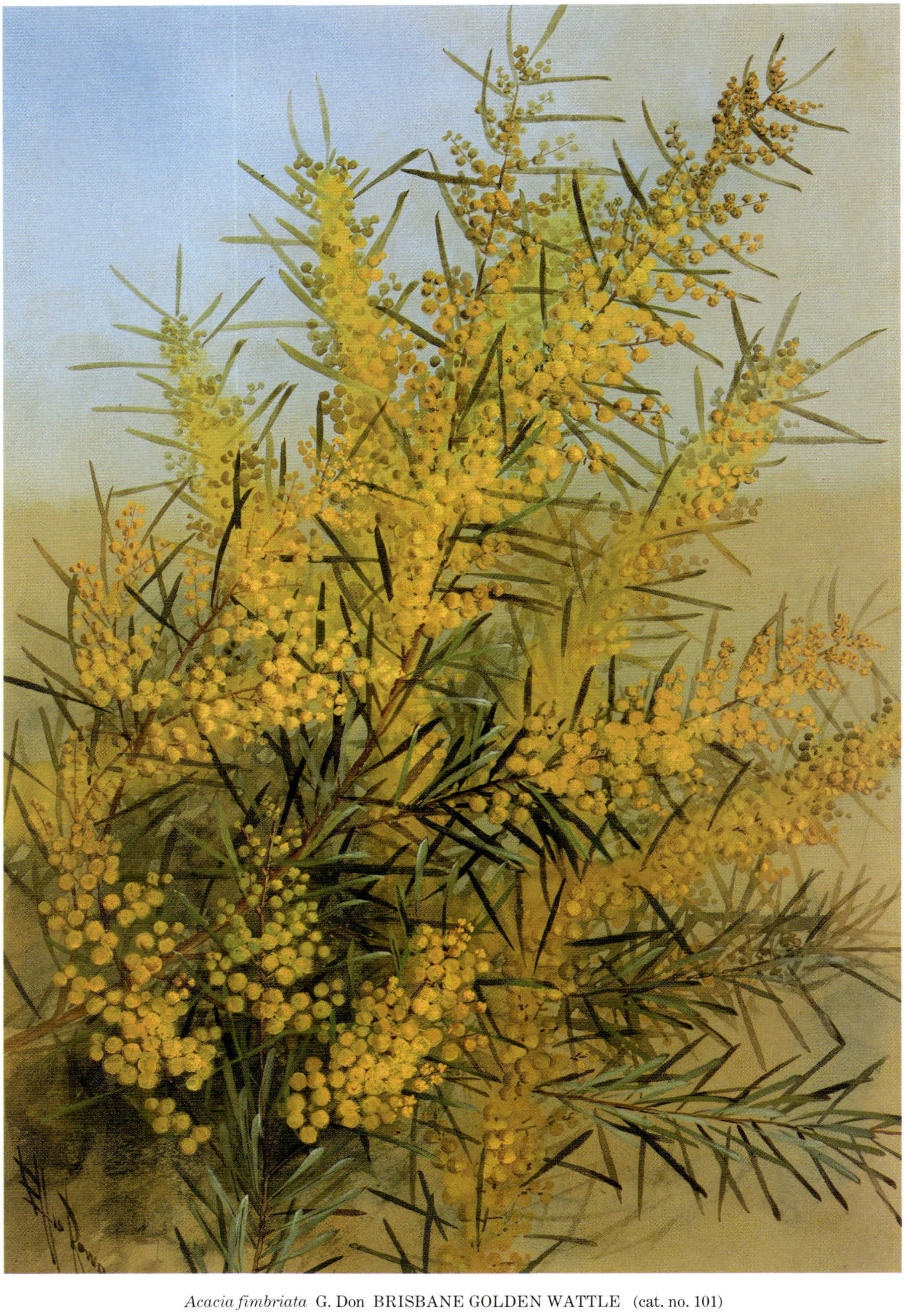

Acacia fimbriata G. Don BRISBANE GOLDEN WATTLE (cat. no. 101)

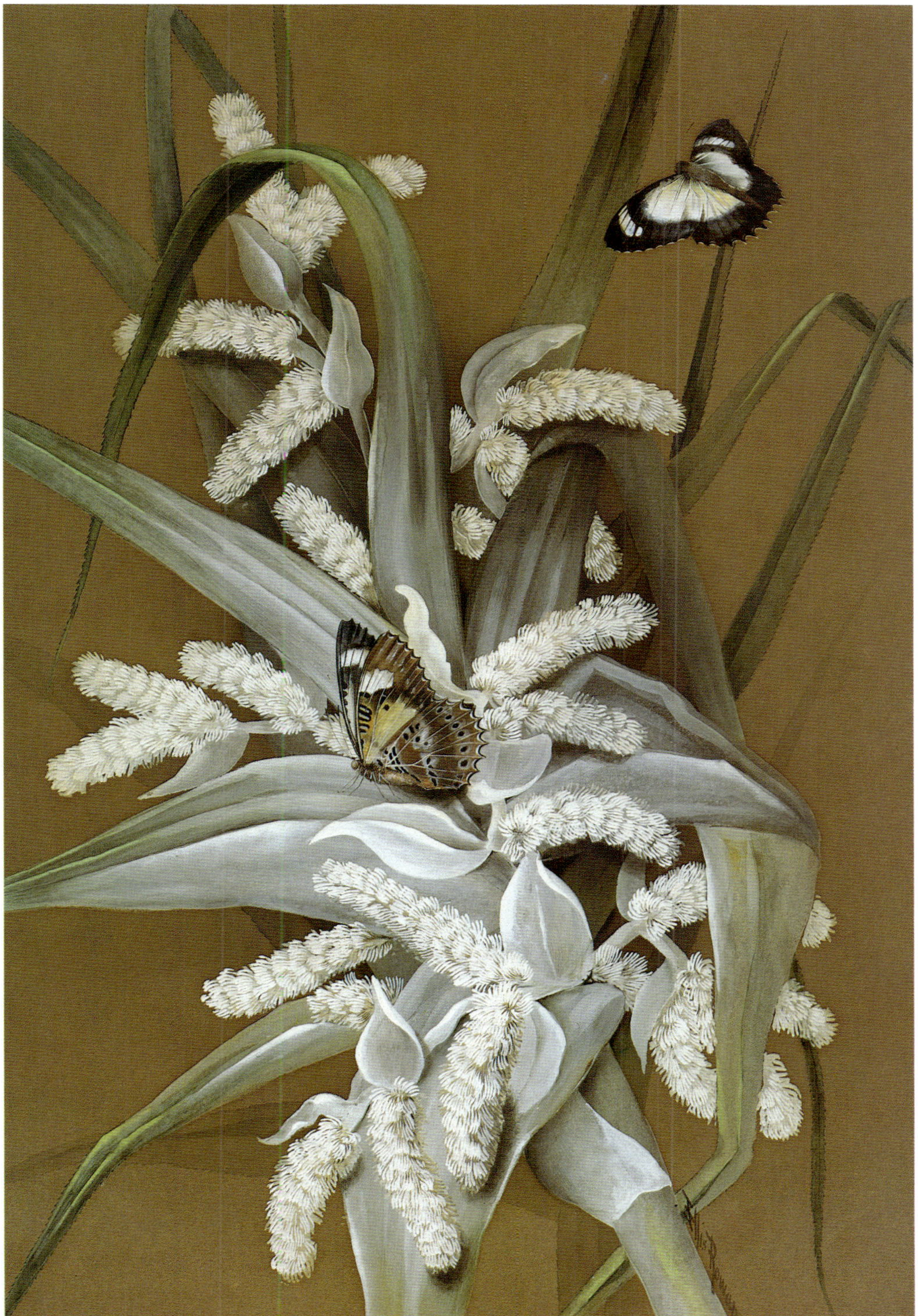

Pandanus monticola F. Muell. SCREWPINE (cat. no. 97)

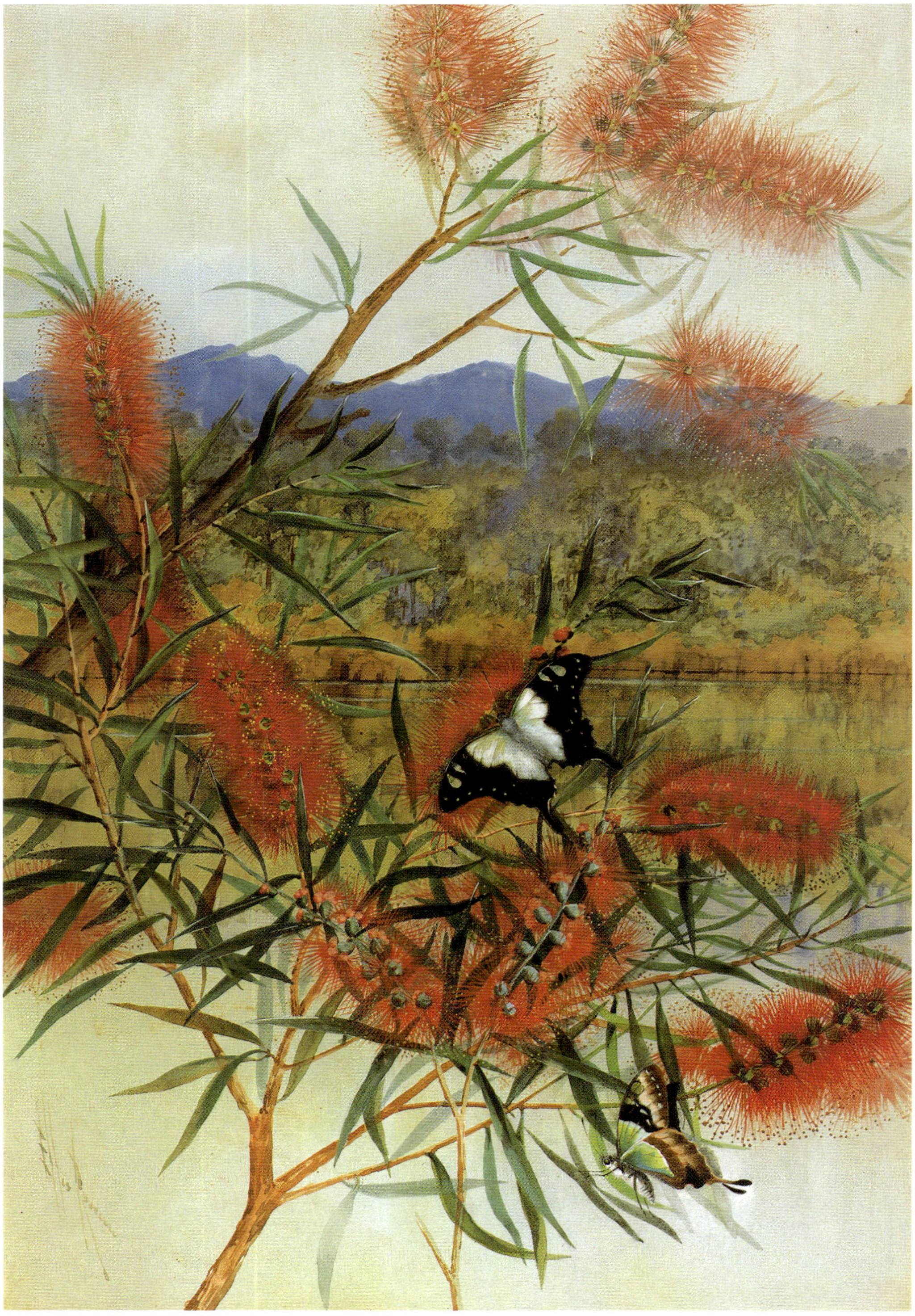

Callistemon viminalis (Gaertner) G. Don ex Loudon WEEPING BOTTLEBRUSH (cat. no. 43)

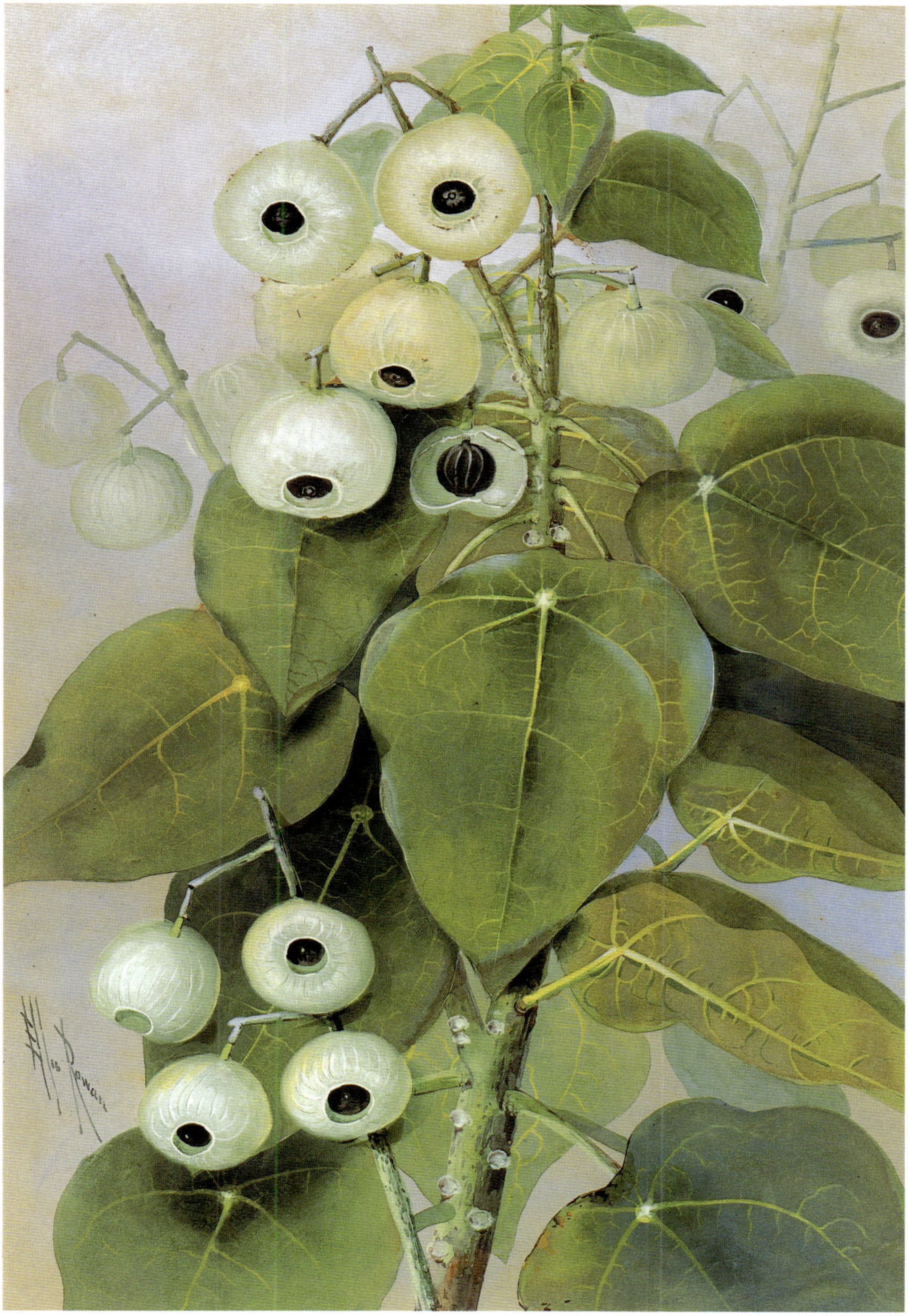

Hernandia peltata Meisner (cat. no. 49)

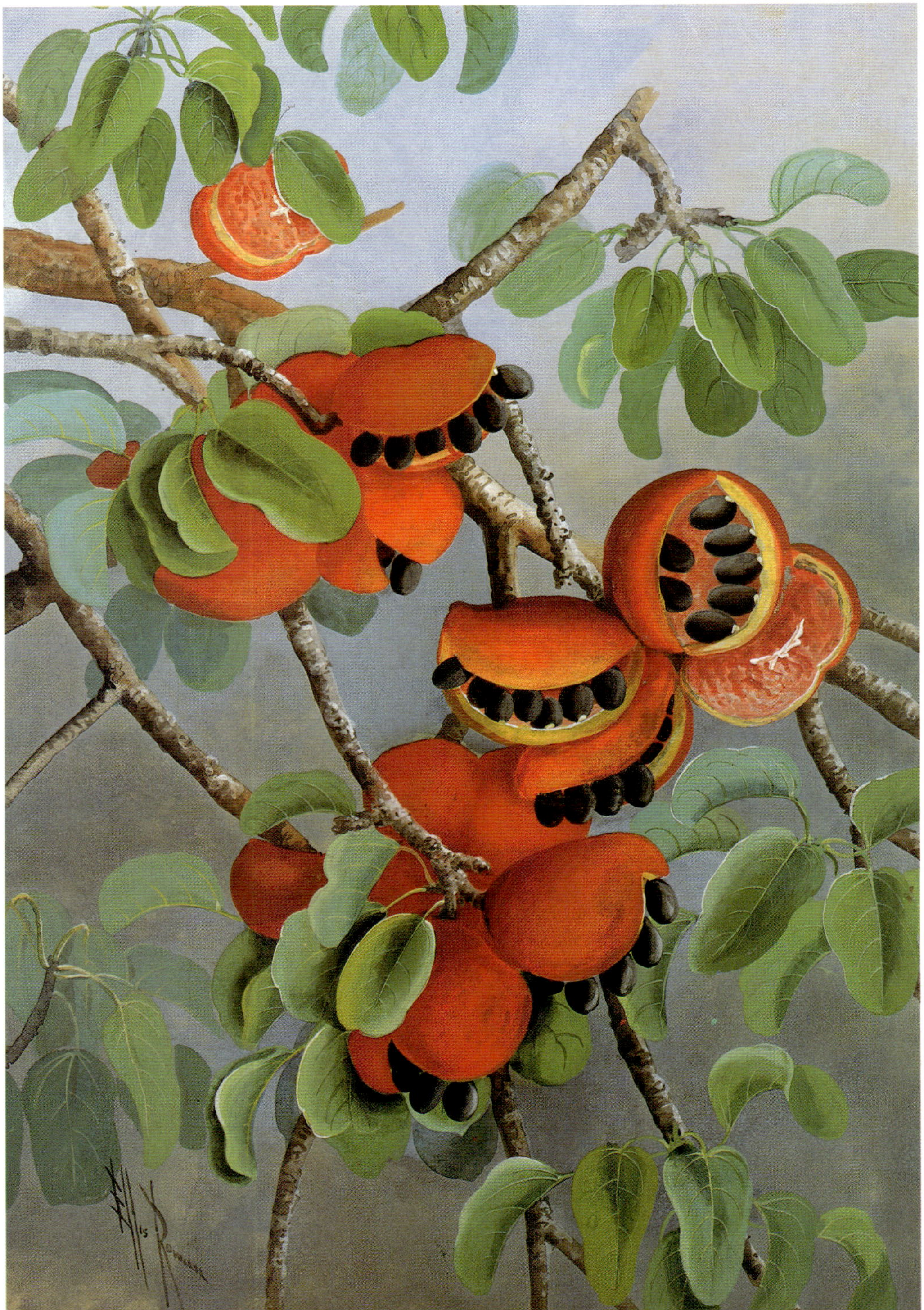

Sterculia quadrifida R. Br. PEANUT TREE; SCARLET–FRUITED KURRAJONG (cat. no. 13)

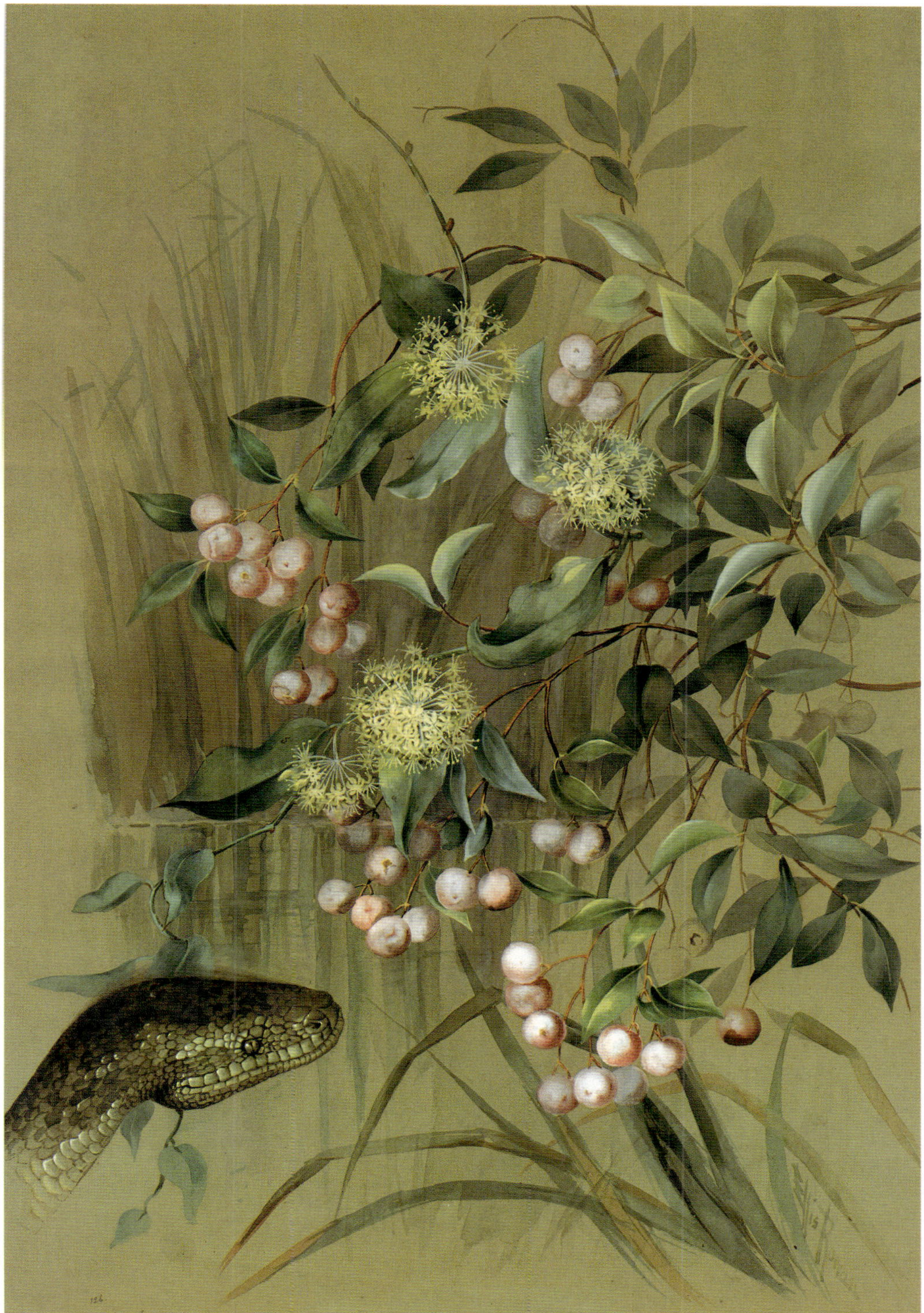

Smilax australis R. Br. BARBWIRE VINE and *Acmena smithii* (Poiret) Merr. & Perry LILLY PILLY (cat. no. 74)

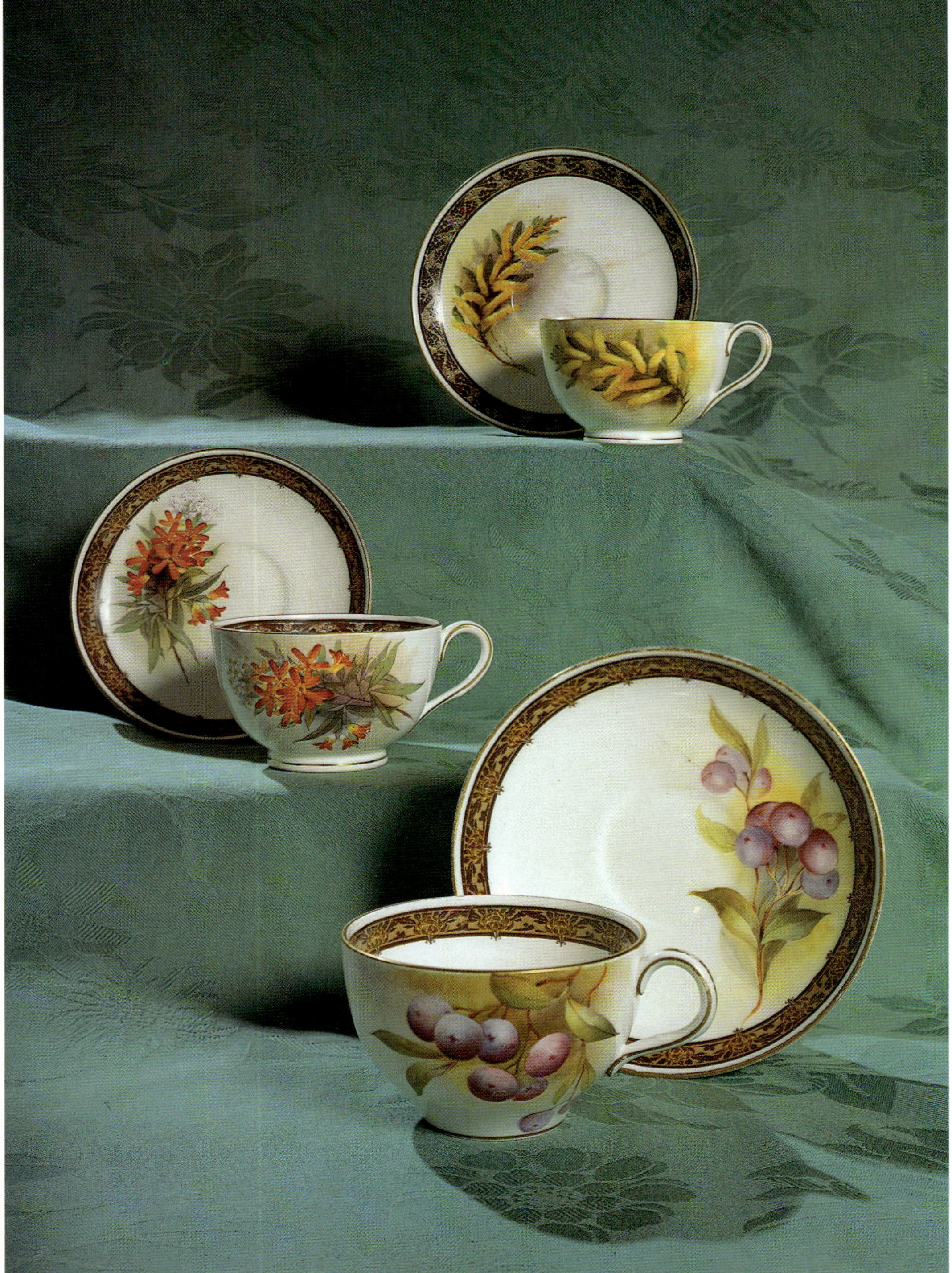

China featuring Ellis Rowan designs — see opposite page

QUEENSLAND MUSEUM COLLECTION

By 19 September 1912, straight after Ellis's victorious exhibition, her collection was taken from the old Town Hall to the Queensland Museum.[1] Under the terms of the purchase, the artist was to "provide a catalogue of the pictures, indicating thereon the names of the flowers and of the localities in which they were painted".[2] Unfortunately, that was a vain hope for Ellis did not keep such records, nor was she always fastidious about plant identifications.[3] Moreover, the paintings were already framed, some many years beforehand for other exhibitions, and any inscriptions on the reverse hidden and forgotten. The "small and meagre"[4] catalogue of her Brisbane exhibition, presumably listing C. T. White's titles, has not survived. Inscriptions revealed by recent conservation treatment have provided new evidence to date and establish localities for many of the paintings.

The paintings were hung along almost the entire southwestern wall of the Museum's upper gallery where, according to Ellis's devotee, the Queensland author and patriot Henry Tardent (1854-1929),[5] they were the best displayed of all Australia's public collections of her paintings:[6]

> Every care has been taken to exhibit those paintings in the best available space in our rather crowded National Museum. They are neatly framed, and shown in two super-posed rows, in the best available light. The great drawback is here the lack of the space necessary to view the paintings in a proper perspective.[7]

In the belief that Ellis was a neglected artist[8] — a view which she herself had promoted — Tardent researched her life and work for a lecture delivered to the Queensland Authors and Artists' Association on 16 March 1926. His grandiloquent lecture was published as a book in the following year, the only book to be published on Ellis for over 50 years.

The upper gallery of the Museum provided less than ideal conditions for the paintings from a conservation viewpoint. Damage from intense light was foreseen in October 1912 when the Director requested blinds for the windows.[9] Water damage presented a more serious problem in 1947 when a storm broke the clerestory windows and water poured down the walls. The paintings were withdrawn to safe storage and displayed again on several occasions in the mid-1960s as backdrops to wild flower exhibitions at the Museum. The first of those exhibitions, in 1964, was organised by the Queensland Naturalists' Club.

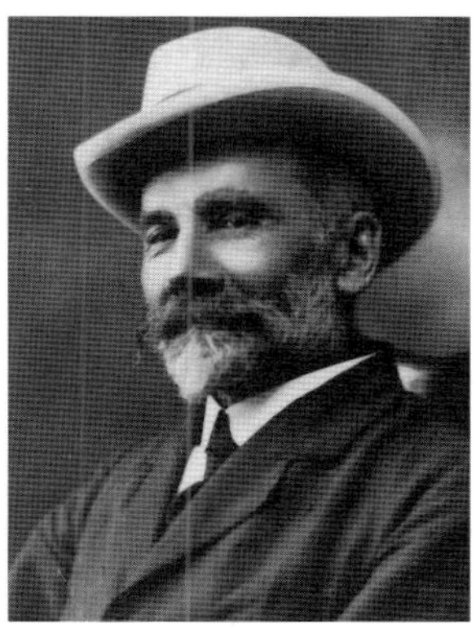

Henry Tardent, a Swiss-born advocate of Australian culture and author of a book on Ellis Rowan.
(Courtesy of Mrs J. Jefferies)

Left: *Examples of china with Ellis Rowan designs, made by the Royal Worcester Porcelain Company of Britain, for the Brisbane jewellers Flavelle, Roberts and Sankey: top and centre from the collection of the Queensland Museum; lower right from the collection of the Brisbane City Hall Art Gallery and Museum and possibly inspired by cat. no. 74 (see p. 55). This cup and saucer, featuring a lilly pilly design, was the gift of Miss Helen Annand of Brisbane, whose artist brother Douglas was commissioned in 1924 by Mr James Sankey to copy some of the Queensland Museum's Ellis Rowan paintings for Royal Worcester china.*

Techniques

Though she always preached Marianne North's dictum that flowers should be recorded on the spot in their native habitats, Ellis in fact did most of her Queensland paintings in the relative comfort of a nearby hotel or homestead. *Flower-Hunter* has many references to how she worked into the night painting flowers collected on excursions [10] or presented to her. [11] Wherever she travelled she became a curiosity and the locals vied to bring her choice and interesting specimens to paint. At Innisfail timber-cutters brought in flowers from the local scrubs. [12] Despite having been painted indoors, the paintings generally have the freshness of works painted in the open air for Ellis was a rapid and direct worker, proud that she could apply the paints without the aid of pencil under-drawing. [13] The paintings generally support her claims:

> I always finish my paintings immediately. If I did not I should not have accomplished what I have. Most of the flowers I have taken to the place where I had been stopping and painted them immediately. I never stop until a painting is finished. Sometimes I have painted all night.[14]

Lady Casey records that Ellis wielded her "forest of brushes" with "ferocious concentration". [15]

Ellis did not pretend to be a serious botanical illustrator, rejecting the scientific demands of that genre. She wrote of her Queensland paintings:

> The collection is painted with a view of showing the general public how the flowers grow with their surroundings — While they are <u>all botanically</u> correct you cannot in the one drawing make a <u>picture</u> of the flower and also show it <u>scientifically</u> drawn in sections which is quite a different study.[16]

Her prime purpose was to make artistic compositions. No doubt she would have been familiar with the most famous of all florilegia of the Victorian era, an extravagant folio volume known as Thornton's *Temple of Flora*, published in serial form between 1799 and 1807. Its publisher, Robert Thornton, engaged leading British artists — not botanical illustrators — to produce "Picturesque Botantical Plates" showing plants in the full richness of their native habitats rather than on conventional plain backgrounds.[17] The *Temple of Flora* inspired traveller-artists like Marianne North who catered for public curiosity in the plant life of distant lands. Likewise in search of the picturesque, Ellis painted Queensland's more spectacular and colourful flora in bold close-up compositions, often extending beyond the picture frame. Some of her close-up views of plants juxtaposed against distant sky (cat. no. 108) or scenery (cat. nos 57, 94) are remarkably similar to the plates in Thornton's pantheon of exotica. Like some of Thornton's artists, she added insects and even snakes for dramatic effect.

Ellis used a highly personalised combination of traditional watercolour techniques and her own ways of drawing detail with her brush in more opaque paints. The naive and lush charm of her paintings gives some credence to her claim to be self-taught. Another of her peculiar and somewhat non-professional practices was to make multiple copies of many of her paintings, presumably her best-sellers. Comparison of similar paintings in the collections of the Queensland Museum and the National Library of Australia suggests that some of the duplicates were done much later, lacking not only von Mueller's botanical notations but also the freshness of the original paintings.[18] Given that Ellis dated only her earlier paintings, the existence of later duplicates adds to the difficulty of classifying her output conservatively estimated at over 3000 paintings.[19]

The paintings resulting from Ellis's various

visits to Queensland may be classified as follows:

1887 paintings: cat. nos 57, 60, 61, 66, 68, 71, 75, 91, 94, 96, 97, 99, 111. These are the only paintings in the Queensland Museum collection to be dated by the artist. Even without the dates, most of the paintings — or "watercolour flower drawings" as they were then called — can be identified by their "Payne's grey" paper and by their technique. With their more contrived and formal plant groupings, finely drawn detail and generally plain backgrounds, these paintings compare with the genteel flower studies of Ellis's painting companion from Western Australia, Margaret Forrest (1844-1929).[20] Even at this stage, however Ellis's paintings are more bold and lush than the dainty flower studies of her Australian female contemporaries, [21] and employ dramatic spotlighting as a reviewer observed:

> The flowers are not only skillfully and faithfully handled but there is always a purpose running through each group. Sometimes in a brilliant cluster of fruit or flowers most of it will be kept well in the shadow and the light only allowed to fall on two or three of the richest specimens. Through this treatment the value of the lighted work is greatly enhanced and made more brilliant still by that which is thrown into shade. [22]

All the original paintings are inscribed in purple pencil on the reverse with a number with a "Q" (for Queensland) prefix and a locality, the numbers following the course of the artist's travels. Curiously, these inscriptions are not in her hand. Most of the paintings also have plant names on the reverse in Ferdinand von Mueller's large distinctive hand. A few had blue crayon numbers on the original mounts, indicating their inclusion in the artist's exhibition at Angus and Robertson's Gallery, Sydney in 1910. All but cat. no. 111 were originally in dark wooden frames with coloured mounts.

1891-92 paintings: cat. nos 22, 41, 42, 46, 51, 54, 56, 58, 59, 63, 67, 70, 72, 74, 76, 77, 80, 82, 83, 84, 85, 86, 88, 89, 90, 92, 93, 95, ?98, 100, 102, 123. These

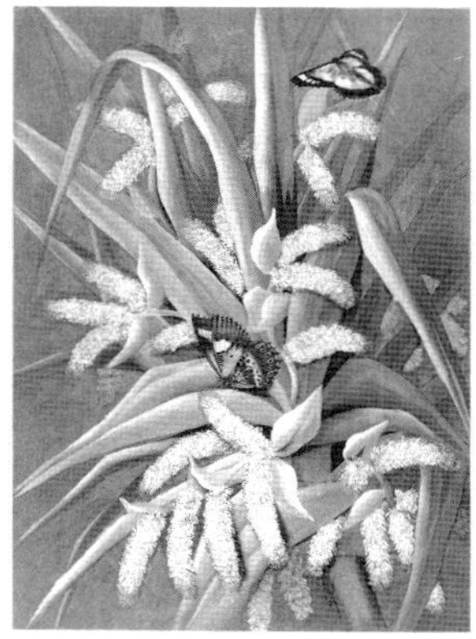

Ellis Rowan's painting of Pandanus monticola F. Muell., a duplicate of cat. no. 97. (Ellis Rowan Collection, R2200, Courtesy of the National Library of Australia)

represent Ellis's mature style which did not vary markedly hereafter. The plants are generally shown more informally in their native habitats, with delicate colour washes replacing plain backgrounds. Green paper is used as commonly as the darker grey paper. Many of the paintings are inscribed on the reverse with a locality and number in the artist's hand and some also with von Mueller's plant names (he died soon afterwards, in 1896). Many have pencil numbers on the lower left corner and or blue crayon numbers on the reverse of the original mounts, again indicating their inclusion in the artist's Sydney exhibition of 1910. All but a few of the paintings were originally in dark wooden frames with coloured mounts.

1911 paintings: the remaining majority of the collection, with the exception of those painted in the southern states and in West Indies. These paintings are generally on green paper, evenly illuminated, and the brushwork occasionally sloppy. Some (notably cat. no. 43) include accomplished landscape habitats. Other close-up compositions (for example, cat. nos. 49, 116) betray some familiarity with Japanese woodcuts, in vogue in the late 19th century. Many of the paintings are inscribed on the reverse with underlined numbers in the artist's hand — in more than one sequence — and with plant names in C. T. White's cursive hand (he was appointed in 1905). Most were originally in gold frames and gold mounts, which were provided by A. L. Hambleton of Brisbane.

Plants *(R. J. F. Henderson, Queensland Herbarium)*

In spite of the assessments by her contemporaries, Ellis Rowan's paintings are of limited value to today's botanists. In varying degrees they lack the critical detail that is necessary and fundamental in so-called "true" botanical illustrations. Nevertheless, the paintings are sufficiently detailed and accurate enough for most of the plants to be identified with little trouble.

Although she was passionate about wild flowers and collected and dried specimens during her Queensland travels,[23] Ellis confessed that she knew "very little" about botany.[24] She seems to have been a trifle too modest. Her accurate naming of strongly scented species in defending the beauty of the Australian bush against claims by writer Marcus Clarke that Australian flowers are scentless, suggests a more than basic knowledge. She wrote:

> To stigmatise them as without scent moreover is a grave injustice, for many of them emit freely a perfume which fills the surrounding air with fragrance. What can be more exquisite or more delicate than the scent of *Boronia megastigma* and *Boronia heterophylla*; of *Boronia serrulata*, Sydney native rose, of many of the acacias; of *Arthropodium strictum*; of *Alyxia buxifolia*, or of the beautiful so-called "rock-lily" of Sydney, *Dendrobium speciosum*?[25]

On the other hand, calling dugongs "graminivorous"[26] (grass-eating) creatures because they browse seagrasses, which are in fact not "true" grasses, seems somewhat amateurish.

Even so, it is astonishing that she was able to produce the botanically accurate paintings she did, especially as there is not any evidence to suggest she used pencil outlines as a guide to proportions of the various parts. Her paintings are a measure of her skill and knowledge, and of the astuteness of her observations as to what the critical parts for identification of the various plants are. No doubt associating with Ferdinand von Mueller gave her some inkling as to what her paintings needed to show to satisfy professional botanists.

In the early stages of her painting career, Ellis seems to have been partly influenced by the then current vogue in flower painting of depicting composites of flowers and / or fruits as isolated bunches or bouquets (cat. no. 75 is typical). After her time spent illustrating for Alice Lounsberry's volumes on American plants, she seems to have depicted plants more in the environment in which they occur, than as studio subjects.

When her paintings are looked at from a scientific point of view, it is clear artistic licence has been taken in several cases to produce aesthetically pleasing depictions. Rather than recording all details of her subjects, background

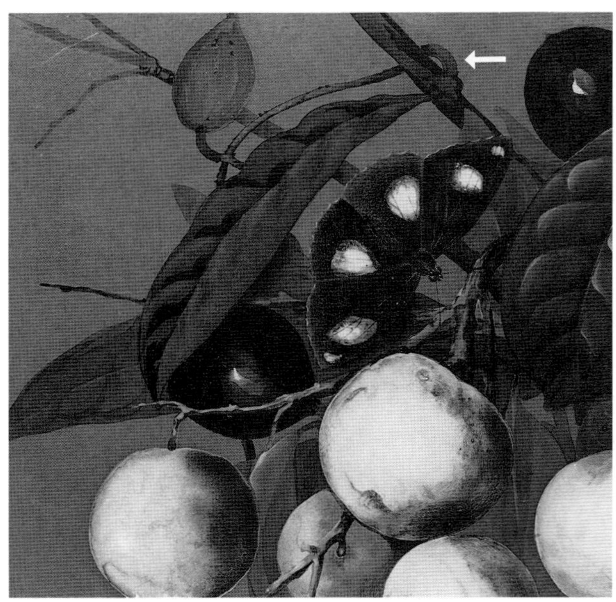

An example of artistic licence — a detail from cat. no. 66 showing the stem of Hoya nicholsoniae emerging from the stem of another plant.

plants and plant associations or landscapes have been rendered anonymous and unobtrusive in some instances (for example, cat. nos 8, 25, 31, 45, 70, 73), while in others she has joined parts of different species on the one stem to produce a pleasing arrangement (cat. no. 54, where a spike of small tubular flowers has been mounted atop a twig of *Lysiphyllum hookeri;* and cat. no. 66, where, in the upper left of the painting above the eggfly butterfly, the stem of *Hoya nicholsoniae* has been made to emerge from the stem of *Niemeyera prunifera*, an entirely unrelated species).

Ellis did not set out to systematically paint the flora of Queensland for its vascular species alone number some 9000 to 10,000. The Queensland Museum collection, itself only a fraction of the many hundreds of Queensland paintings which she exhibited and sold over several decades, represents only a small section of Queensland's flora. She apparently preferred the more spectacular and colourful species and included several cultivated garden plants in her paintings. The stations and sugar plantations where she stayed in Queensland generally had well-stocked fruit and flower gardens with a range of exotic species,[27] a fertile source of subjects to paint. The plants she depicted tend to be more showy species, many of which are trees or shrubs, though conspicuously flowered vines are also well represented.

It seems the many small-flowered herbaceous species of our flora were not among her favourite plants. She does appear, however, to have had a special love of waterlilies for she comments repeatedly on them throughout *Flower-Hunter* and included three paintings of them in her Brisbane selection (cat. nos 57, 64 121), all duplicated in the National Library collection.

Nevertheless, the Queensland Museum paintings cover a fair range of species diversity despite the exotics and the six species illustrated more than once or twice (for example, *Cassia queenslandica* cat. nos 22, 26, 62, and *Melaleuca quinquenervia* cat. nos 29, 32, 35). The economically

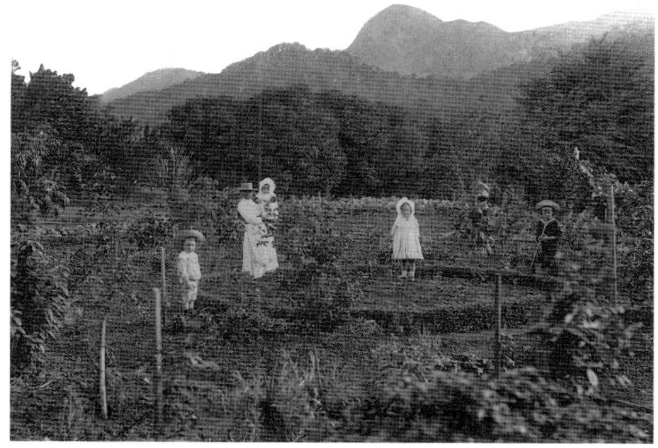

The garden at Hambledon plantation, which grew tropical fruits, peanuts and ginger. Ellis had her first taste of custard apples there. (John Oxley Library, *19411*)

important legume families Fabaceae, Caesalpiniaceae and Mimosaceae are certainly well represented, as is the cotton family Malvaceae and the myrtle family Myrtaceae. Depictions from the important grass family Poaceae do not appear as often, no doubt due to the lack of showiness among its members. The large and important sunflower (daisy) family Asteraceae is not represented at all, perhaps because none of its species took her fancy.

In making her selection for her Brisbane exhibition, Ellis was not concerned about confining the paintings to Queensland's flora. She included exotic garden-plants (cat. nos 27, 30 painted in Queensland, and 42 painted in New Guinea), and even some not known to occur naturally in Queensland (cat. nos 79, 81, 87) or Australia (cat. no. 65).

Though the scientific value of the collection is limited, it is an outstanding historical record of some of Queensland's flora painted by a botanically aware artist in search of the picturesque.

Animals *(G. B. Monteith, Queensland Museum)*

Ellis Rowan is known as a flower painter, a description she bestowed upon herself, but she was far from being a painter of only flowers and plants. Her major ventures into animal painting were her series of New Guinea bird and butterfly paintings completed in her later career.[28] However in the 125 paintings held by the Queensland Museum we see animal painting of a more subtle kind because one-third of them (forty-one) contain drawings of smaller animals, mostly colourful insects, added to complement and enliven the main plant subject matter. The thirty-two species depicted (a number are repeated) cover a great variety of creatures, including two dragonflies, a grasshopper, a mantid, a beetle, a shield bug, a fly, five moths, sixteen butterflies, two spiders and two snakes.

We know from her own accounts, and from the observations of her contemporaries, that her flower painting was done mostly indoors, often at night, and from plant specimens picked by herself or brought to her by local people. It is clear that the included animals were not painted *in situ* but must have been added from a separate subject source and, perhaps, at a later time. This being the case it is worth considering whether these "after-thought" animals are, in fact, accurate representations of real species, and if so, whether they are shown in correct contextual association. In other words, were her animals ones which she herself saw at the time she collected the plants, or were they just imaginary animals, or animals copied later from specimens and books, and added for purely artistic whim? Richard Semon, a German biologist who met Ellis in Cooktown in 1892, has already made a judgement in this respect:

> Her pictures were perfect works of art, combining scientific exactness with aesthetic beauty! I had never before seen botanic studies of similar perfection. To enliven her pictures, she had here and there put in a handsome Australian insect, beetle, or butterfly, sitting on the plants, or flitting around them. It was a pity, however, that she had grouped plants and insects at random, without caring if any relationship existed between them. This proceeding, though indisputable from an artistic point of view, seems to me not quite fortunate in paintings principally destined for scientific use.[29]

Ellis Rowan had a natural affinity for animals acquired from her early family contacts, as recollected by her niece Lady Casey:

> She had inherited the interest of her grandfather John Cotton not only in flora and fauna but in insects and even reptiles; she enjoyed the accidents of life, the fusion of one aspect of nature with another. ... Some of our family, the Le Souefs, actually had a feeling for snakes. They liked running their warm fingers over those long cold exquisite bodies.[30]

This background, and her artistic motivations, translated into a keenly observant eye for the smaller animal life during her painting expeditions. Indeed, in *Flower-Hunter* her lively descriptions of encounters with butterflies, mantids, dragonflies and so on usually overshadow her references to plants which are often relegated to tales of marathon painting sessions. One gains the impression that the insects gave her entertainment and relief from the business of flower painting.

Ellis was not only an observer, but also a collector of insects, and she travelled equipped for this. Her letters often refer to collecting activities. At Macknade: "a lovely pale green moth ... was quickly caught, chloroformed and pinned out".[31] At Goondi:

> I handed my net over to a small black boy, who, instead of butterflies, brought me a fat green

grasshopper, well squashed in his moist little hands! ... I spent hours in hunting butterflies, which are found there in wonderful variety ... one (moth) that I caught I have since heard is a new variety ... [32]

Some of these specimens she was painting during her travels, for at Habana she writes: "I have already painted one or two very large Praying Mantis".[33] Ellis's letters are replete with adventurous stories of snakes and some also became immediate subjects for her brush. At Mackay in 1887 she sketched a bird-eating snake at night, then after it was despatched by her host Mr Rawson — "Next day I filled in the sketch in colour, getting also correctly the scales ..." [34] At a "naturalist's establishment" in Cairns she painted "a small, very lively black snake with the figure 2 in white most perfectly shaped in the middle of its head".[35]

Ellis claimed that two specimens she collected — a snake and a moth — eventually found their way to the British Museum as possible new species.[36] The fate of the other specimens collected is uncertain, but a grand-niece had childhood memories of a visit to her attic studio where "there were sheets of paper leaning against the wall at the back of a shelf with rows and rows of butterflies painted on them ..."[37] This perhaps indicates the preparation of reference studies of butterflies as a source for adding to her flower paintings. Comparison with the Ellis Rowan Collection in the National Library shows that she added and deleted insects from her duplicate paintings as fancy took her.[38]

With this sketchy background of what we know of her *modus operandi* in tackling the "afterthought animals" of her flower paintings we can assess the scientific precision with which she illustrated them in the Queensland Museum paintings.

Certainly all the animals in her paintings are real species and almost all can be readily identified from her depictions (see Catalogue pp. 65-81). The only evidence of contrived insects are rather unconvincingly anonymous beetles in cat. no. 105

Ellis Rowan's painting of a brown tree snake "with a bird just caught between its jaws" and Dendrobium smillieae, painted from life at Mackay in 1887. (Ellis Rowan Collection, R2195, courtesy of National Library of Australia)

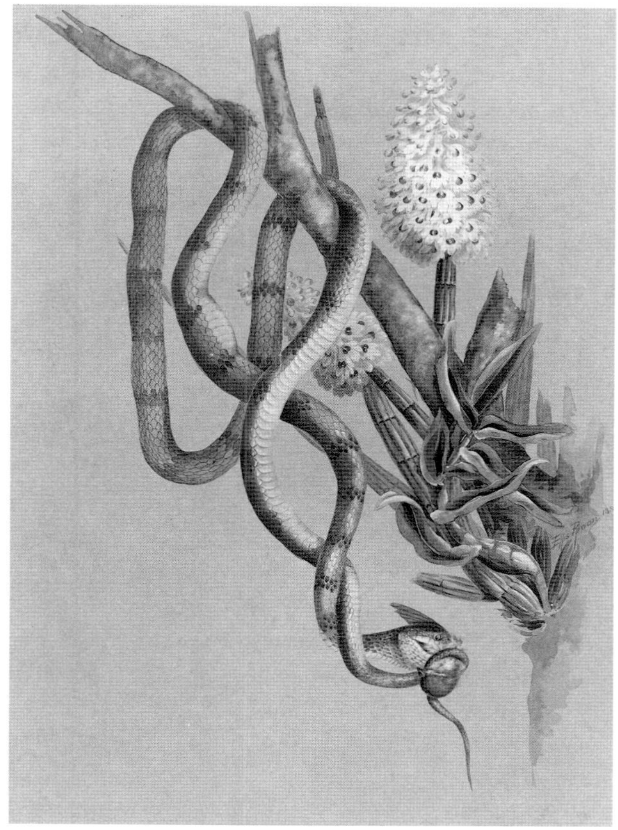

and the grasshopper in cat. no. 28. A degree of artistic licence is evident in a few paintings. For example, the bright red patches on the jezabel butterfly *Delias mysis* (one of her favourites, appearing in three paintings, cat. nos 10, 79, 92) are abnormally enlarged, while the brilliant red in the *Cethosia cydippe* (cat. no. 97) is absent, perhaps to blend it with the large white inflorescence of *Pandanus monticola* which dominates the painting; the same butterfly is correctly coloured in cat. no. 40. Certain butterflies are shown with incorrect wingshape (for example, *Hypolimnas bolina* in cat. no. 66) perhaps evidence of having been copies from earlier paintings rather than from life.

Finally we can consider the naturalness of the biological setting of the animals in the paintings. The three aspects which contribute to this reality are, firstly, whether they are shown in life-like attitudes, secondly, whether the animals genuinely associate with the plant shown, and thirdly, whether the geographic range of the animal actually coincides with that of the plant. Ellis was a keen observer of butterflies in life and this comes through strongly in her paintings which show them in many complex attitudes of perching and flight which are quite accurate (for example, cat. nos 7, 24, 63, 68, 97). This is not the case with the two spider paintings (cat. nos 29, 30) which show normally web-resting *Nephila* spiders in unnatural postures on foliage. The two moths in cat. nos 9 and 37 are clearly based on dead specimens, the former having curled antennae and part-spread wings, the latter having legs in the attitude of *rigor mortis*. But the dominant effect is of animals in natural, often dynamic, poses.

Most butterflies and other insects have associations with particular plant species on which their caterpillars breed. Though Ellis loved butterflies she is on record as asserting that "caterpillars are the things my soul abhors".[39] So perhaps it is not surprising that Richard Semon's criticism of her random association of animals and plants is borne out by this collection of her Queensland paintings. Clearly she either knew not or cared not where her butterflies planned to deposit their "abhorrent" caterpillars because only one painting (cat. no. 124) shows an insect (*Anaphaeis java*) with a plant (*Capparis spinosa* var. *nummularia*) on which it might breed. This supports the view that the animals were added to the paintings largely for artistic effect.

The geographic context of the painted animals appears accurate in all cases with the exceptions of cat. nos 79 and 92. All the animals occur in Queensland and within the range of the plants with which they are shown. In cat. no. 92 the butterfly *Delias mysis* rarely occurs as far south as Yeppoon while the plant *Melaleuca sieberi* occurs only as far north as Bundaberg. This leaves a hiatus of 250 km between the natural ranges of the butterfly and the plant and shows she could not have seen them together. In cat. no. 79 this same butterfly species is shown in a painting done in Victoria and including a plant species not occurring north of New South Wales.

The animals which Ellis included in her Queensland paintings add life, colour and interest to them. The overall impression gained from this analysis is that the species she utilized were real species which she had actually encountered in the same general area from which she obtained her plant subjects, but that they were added to individual paintings under the exigencies of artistic whim.

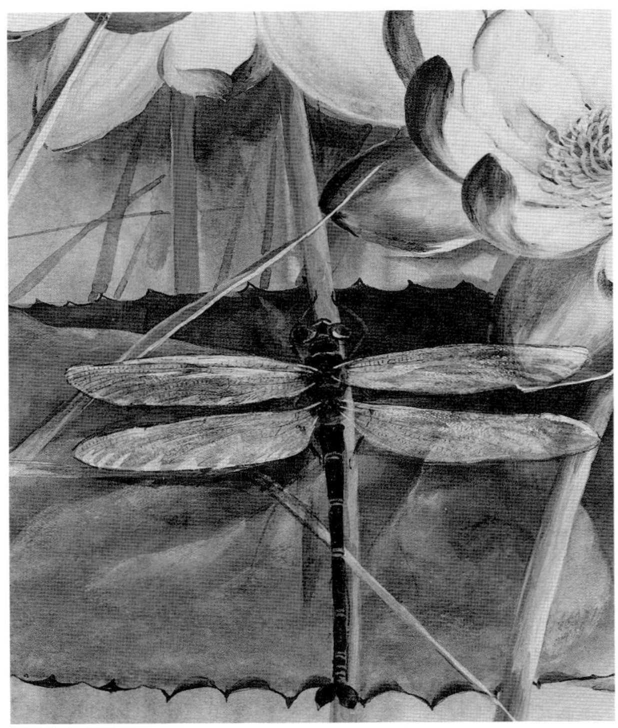

Dragonflies often caught Ellis Rowan's attention. This detail from cat. no. 64 is accurate enough to identify the species as Petalura ingentissima Tillyard, the largest dragonfly in Australia.

Catalogue

Plant identifications were checked and updated in November 1989 by Rod Henderson, Supervising Botanist, and colleagues of the Queensland Herbarium. The current names differ in many cases from the original titles of the paintings. Common names, plant distributions and indexes were provided in 1989 by Queensland Museum volunteers, John and Margaret Parham. Animal identifications were provided in 1989 by Geoff Monteith, Senior Curator Lower Entomology, of the Queensland Museum.

Common names of plants and animals are given in upper case; family names are in square brackets.

Catalogue numbers are those first allocated after the collection was acquired by the Museum.

The artist has signed all the paintings with the exceptions of cat. nos 56 and 125 which may be unfinished or faded.

She has dated relatively few of the paintings — only those from her 1887 visit. Dates established by intrinsic evidence (inscriptions) and / or extrinsic evidence (exhibitions or related paintings in the NLA collection) are given in brackets. However, these dates may change in light of new evidence.

All the works are painted in watercolour and gouache on toned wove paper.

All are of fairly uniform size, the paper ranging from 52.8 cm to 56.4 cm in height and from 37.6 cm to 38.5 cm in width. The 19th century paintings are generally the smaller.

Exhibitions listed are in addition to the artist's Brisbane exhibition of 1912. However, these listings are not comprehensive since some of the relevant exhibition catalogues have not survived and, of those that do survive, painting titles are often not explicit enough to enable identification. Likewise, it is not possible to determine whether original or duplicate paintings were exhibited in the earlier exhibitions.

The following abbreviations are used:

Flower-Hunter	Ellis Rowan, *A Flower-Hunter in Queensland and New Zealand*, 1898;
Bailey	F. Manson Bailey, *Comprehensive Catalogue of Queensland Plants*, (1913);
Tardent	Henry A. Tardent, *Mrs Ellis Rowan and her Contributions to Australian Art and Science*, 1927;
Melbourne 1888	Centennial International Exhibition, Melbourne, 1888;
Chicago 1893	World's Columbian Exposition, Chicago, 1893;
London 1896	Ellis Rowan exhibition, Dowdeswell Galleries, London, 1896;
Melbourne 1904	Ellis Rowan exhibition, Scourfield Chambers, Melbourne, 1904;
Sydney 1910	Ellis Rowan exhibition, Angus and Robertson's Gallery, Sydney, 1910.

1

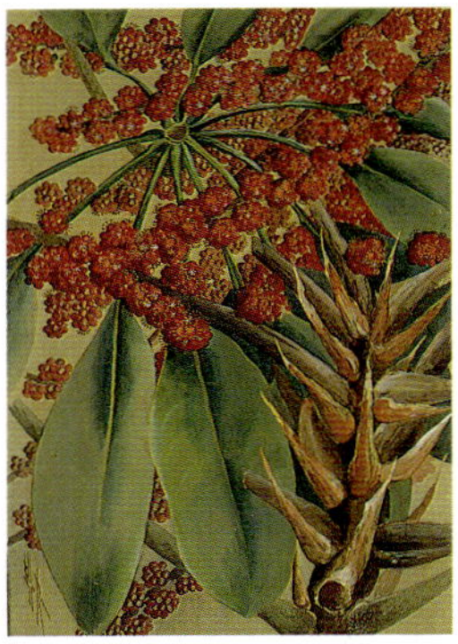

2

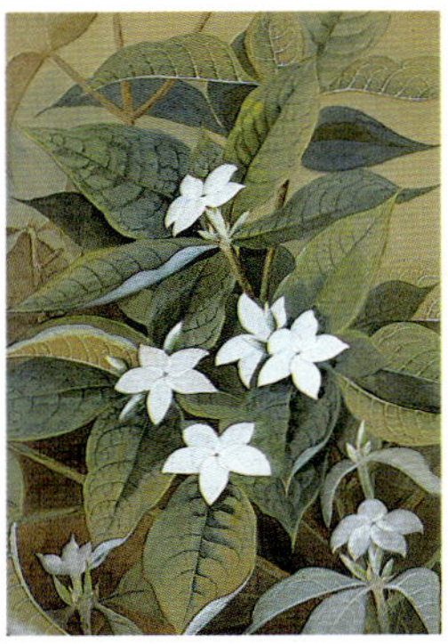

3

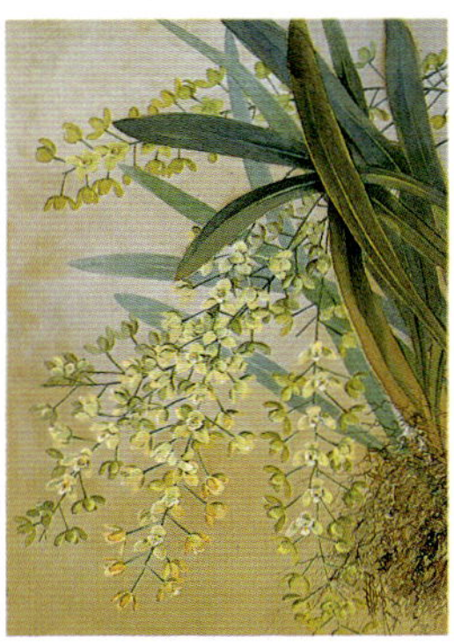

4

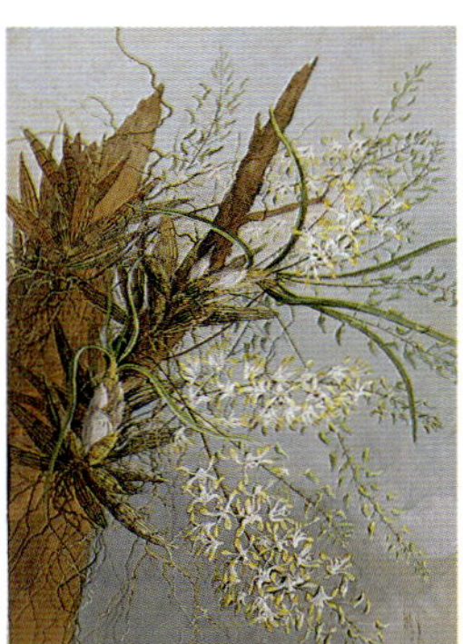

5

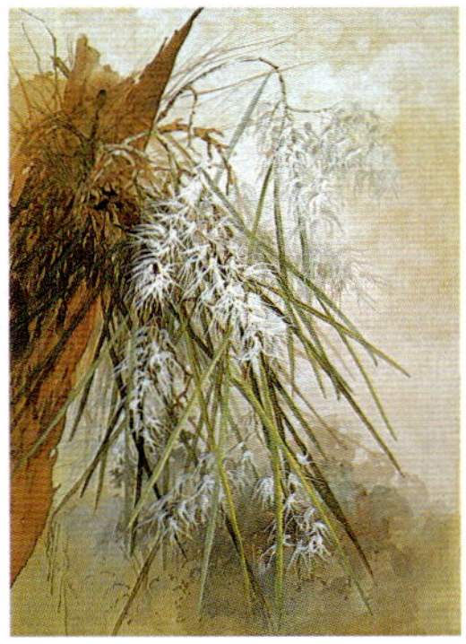

6

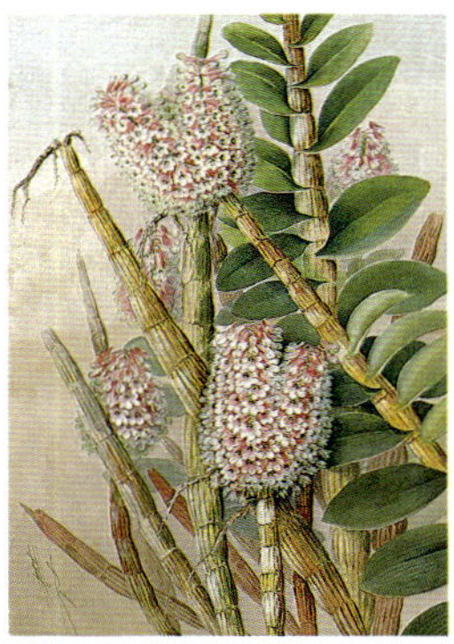

7

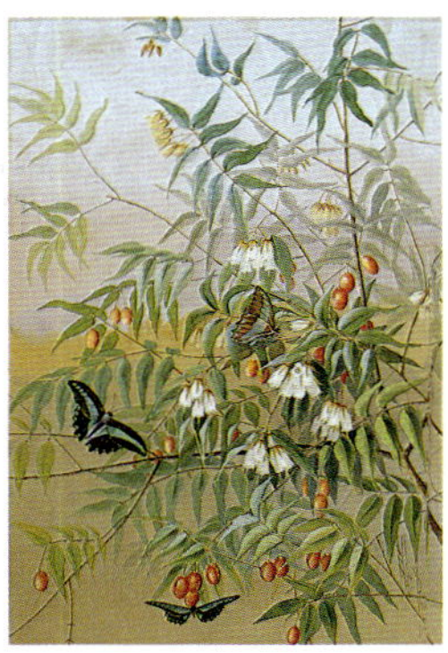

8

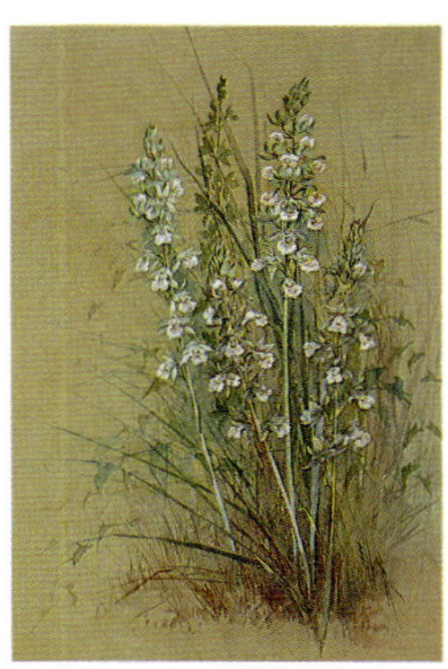

9

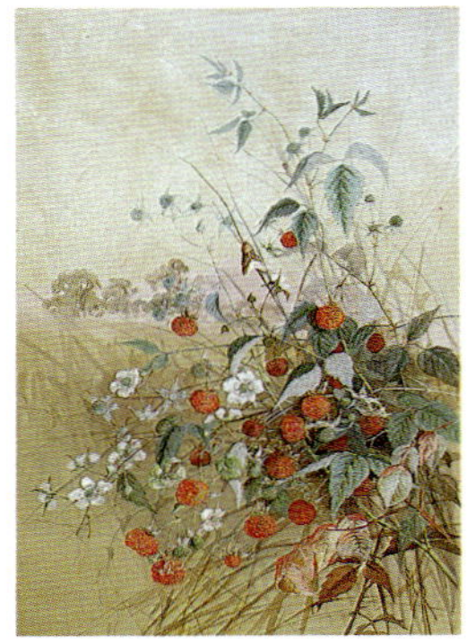

10

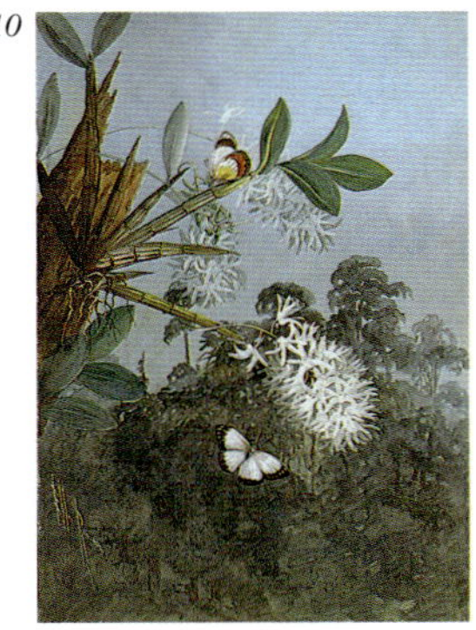

11

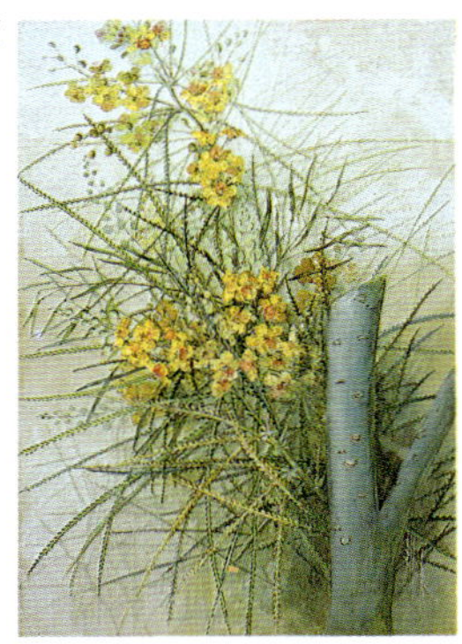

12

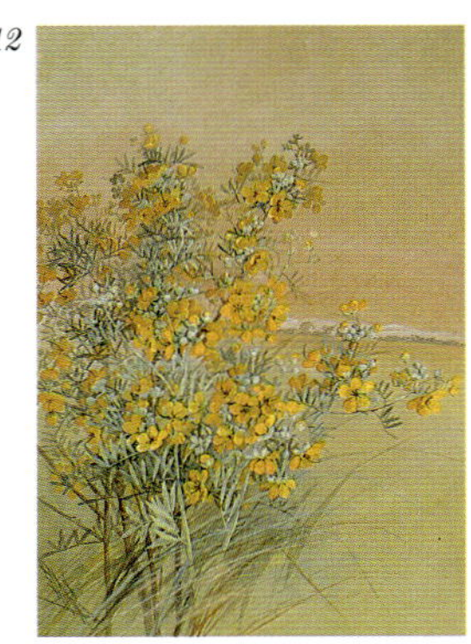

13

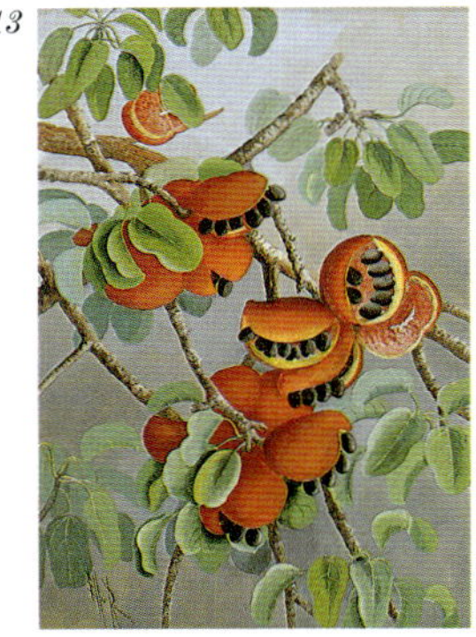

14

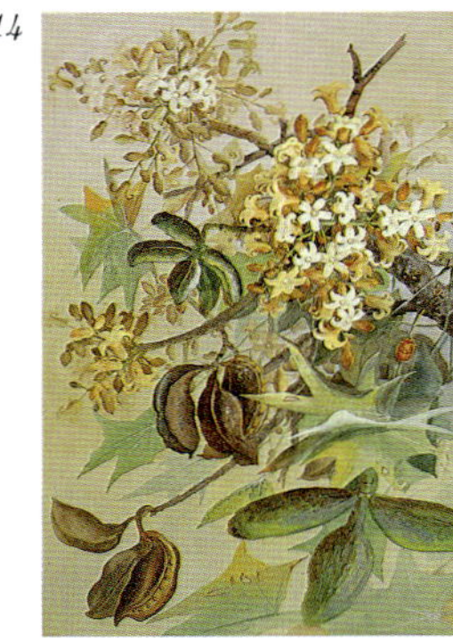

15

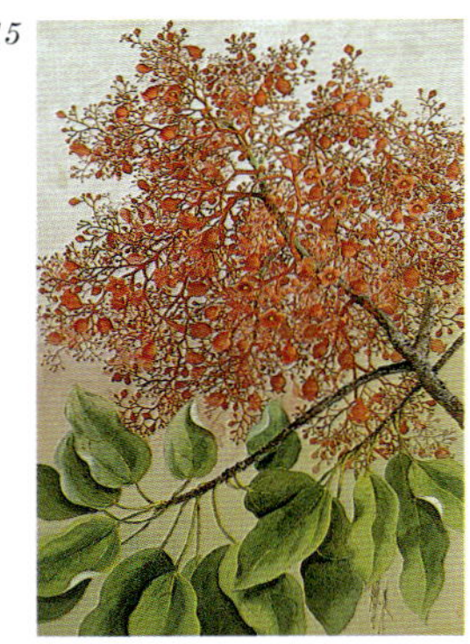

16

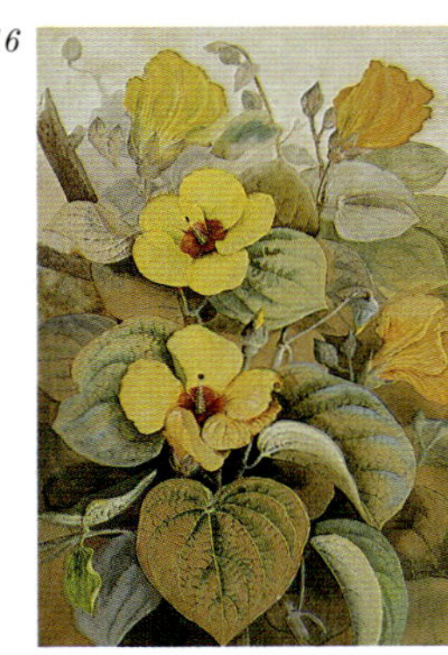

1
Schefflera actinophylla (Endl.) Harms [Araliaceae]
UMBRELLA TREE
Rainforests of Qld N of Mackay; NT; widely cultivated

Painted: (1911)

Numbered by artist reverse "55"
Titled by C.T. White reverse "Umbrella Tree / Brassaia / actinophylla"

2
Randia hirta F.Muell. [Rubiaceae]
NATIVE LOQUAT
Rainforests of Qld N from Cairns area

Painted: (1911)

Numbered by artist reverse "20"
Titled by C.T. White reverse "Gardenia sp."

3
Cymbidium madidum Lindley [Orchidaceae]
E coast of Qld; NSW

Painted: (1911)

Numbered by artist reverse "5"
Titled by C.T. White reverse "Cymbidium sp."

4
Dendrobium canaliculatum R.Br. [Orchidaceae]
TEA-TREE ORCHID
Tea-tree swamps of Qld N from Mackay area

Painted: (1911)

Numbered by artist reverse "5"
Titled by C.T. White reverse "Dendrobium / canaliculatum"

5
Dendrobium teretifolium R.Br. [Orchidaceae]
PENCIL ORCHID
Mainly rainforests of Qld; NSW; New Guinea and other SW Pacific Is.

Painted: (1911)

Numbered by artist reverse "13"
Titled by C.T. White reverse "Dendrobium / teretifolium"

6
Dendrobium smillieae F.Muell. [Orchidaceae]
BOTTLEBRUSH ORCHID
E coast of Qld N from Townsville area

Painted: (1911)

Numbered by artist reverse "27"
Titled by C.T. White reverse "Dendrobium / Smilliae"

7
Aceratium megalospermum (F.Muell.) Balg. [Elaeocarpaceae]
Rainforests of Qld N of Townsville

Insects—*Graphium sarpedon* L. [Papilionidae]
BLUE TRIANGLE BUTTERFLY

Painted: (1911)

8
Flower stems—*Eulophia fitzalanii* Benth. [Orchidaceae]
E coast of Qld

Background—Poaceae (grass) and Convolvulaceae (twiner)

Painted: (1911)

Numbered by artist reverse "38"
Titled by C.T. White reverse "Eulophia sp."

9
Rubus rosifolius Smith [Rosaceae]
NATIVE RASPBERRY
SE Qld; NSW, Vic.; New Guinea, SE Asia, Philippines, Indonesia

Insect—*Spodoptera picta* (Guérin-Méneville) [Noctuidae]
LILY MOTH

Painted: (1911)

Numbered by artist reverse "9"
Titled by C.T. White reverse "Rubus / rosaefolius"

10
Dendrobium ruppianum A. Hawkes [Orchidaceae]
E coast of Qld N of Townsville; NSW

Insects—*Delias mysis* (Fabricius) [Pieridae]
UNION JACK BUTTERFLY

Painted: (1911)

Numbered by artist reverse "401"
Titled by C.T. White reverse "Dendrobium / fusiforme"

11
Parkinsonia aculeata L. [Caesalpiniaceae]
JERUSALEM THORN
Introduced to Australia from S America and widely naturalised in NE

Painted: (1911)

Numbered by artist reverse "19"
Titled by C.T. White reverse "Parkinsonia / aculeata"

12
Above—*Cassia surattensis* N. Burman [Caesalpiniaceae]
E coast of Qld S from Townsville area; NSW, NT, WA

Base—Poaceae (grass)

Painted: (1911)

Numbered by artist reverse "43"
Titled by C.T. White reverse "Cassia / glauca"

13
Sterculia quadrifida R.Br. [Sterculiaceae]
PEANUT TREE; SCARLET-FRUITED KURRAJONG
Rainforests of Qld; NSW, NT, WA

Painted: (1911)

Numbered by artist reverse "16"
Titled by C.T. White reverse "Sterculia / quadrifida"
Inscribed reverse "Mrs Rowan / 16 Brown and / Crown Sts / Be particular to put / correct number on each"

Similar work reproduced: Bailey, plate I; Tardent, p. 45; *Muses Magazine*, December 1927

14
Brachychiton australis (Schott & Endl.) A. Terracc. [Sterculiaceae]
BROAD-LEAVED BOTTLETREE
NE and C coast of Qld

Insect—*Tectocoris diophthalmus* (Thunberg) [Scutelleridae]
HARLEQUIN BUG

Painted: (1911)

Numbered by artist reverse "62"
Titled by C.T. White reverse "Broad leaved / Bottle Tree / Sterculia / Trichosiphon"

Similar work reproduced: Bailey, plate 3; Tardent, p. 25; *Muses Magazine*, March 1928

15
Brachychiton acerifolius (Cunn. ex G.Don) Macarthur [Sterculiaceae]
FLAME TREE; FLAME KURRAJONG
Rainforests of Qld; NSW

Painted: (1911)

16
Hibiscus tiliaceus L. [Malvaceae]
COTTON TREE; COTTONWOOD
Coastal Qld; NSW, NT; Pantropical

Painted: (1911)

Numbered by artist reverse "45"
Titled by C.T. White reverse "Hibiscus / tiliaceus"

17

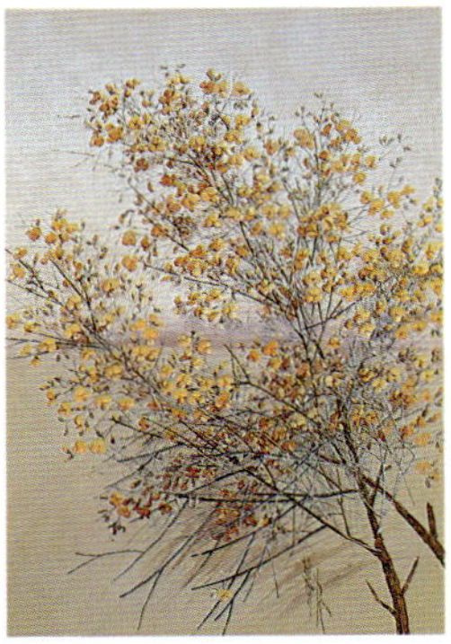

18

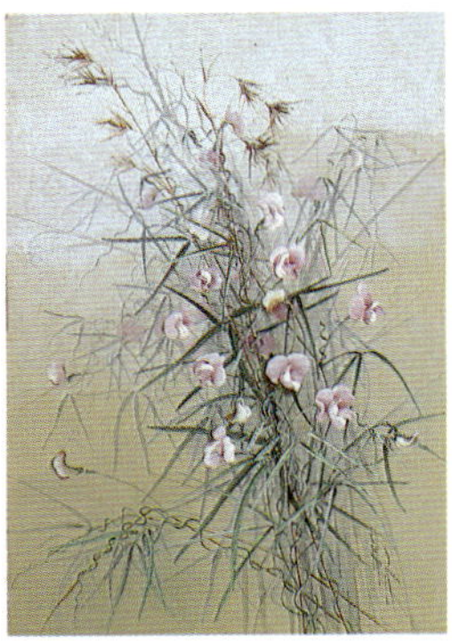

19

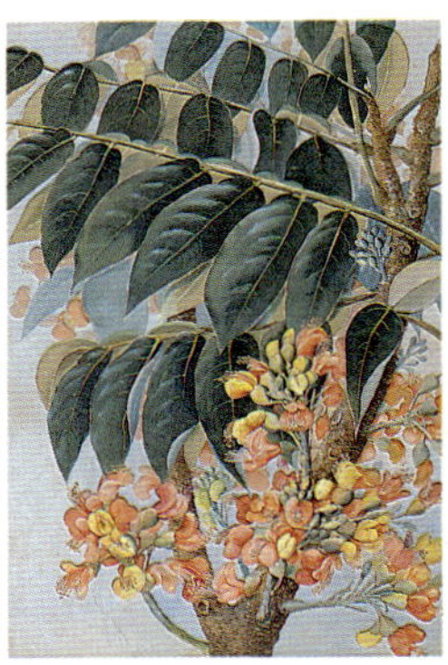

20

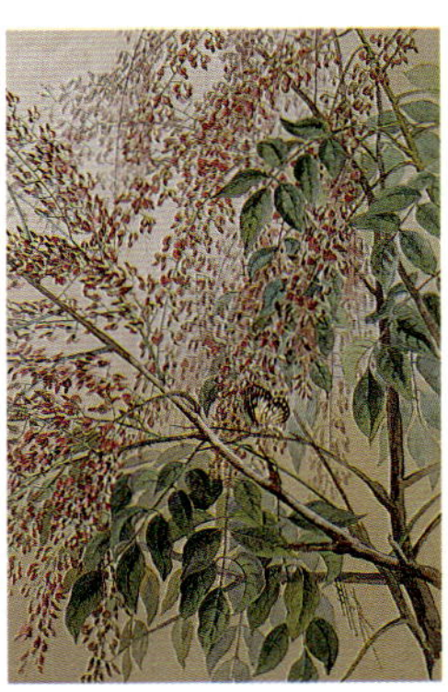

21

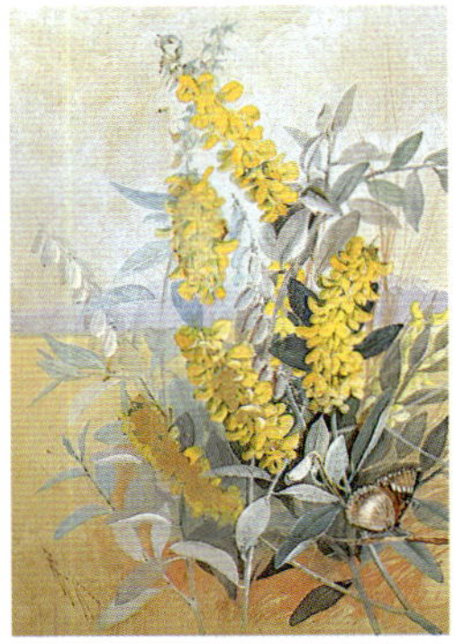

22

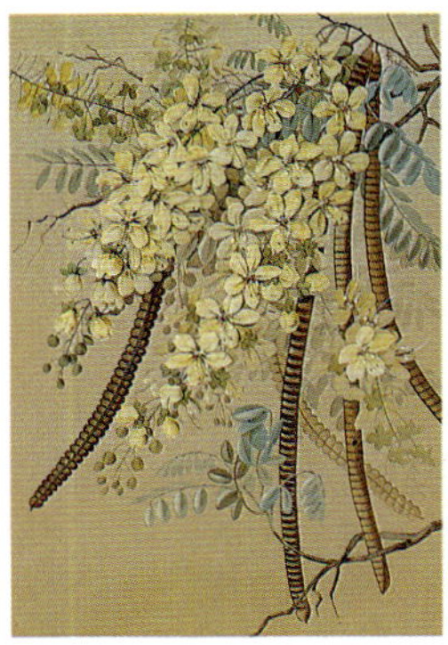

23

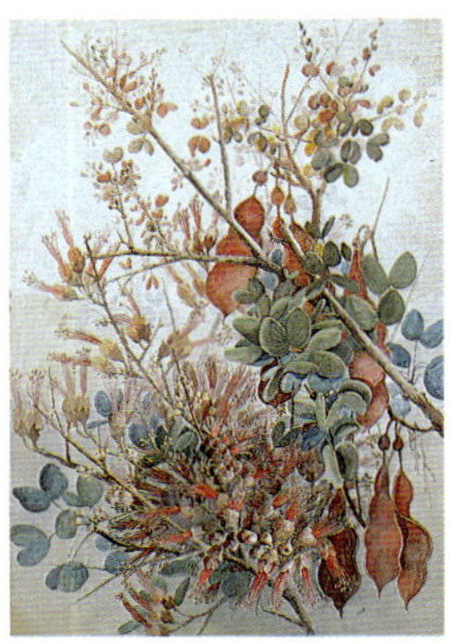

24

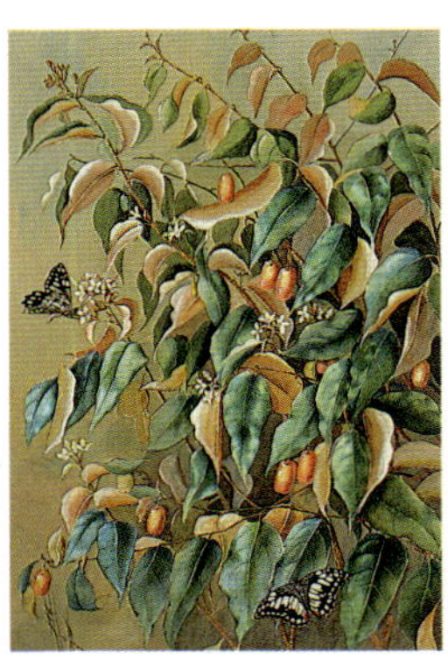

25

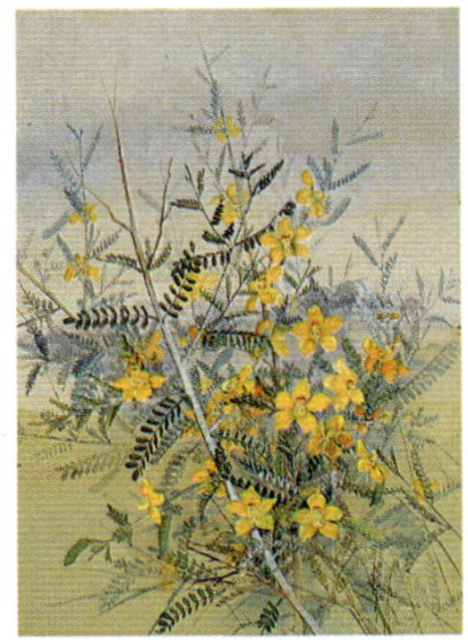

26

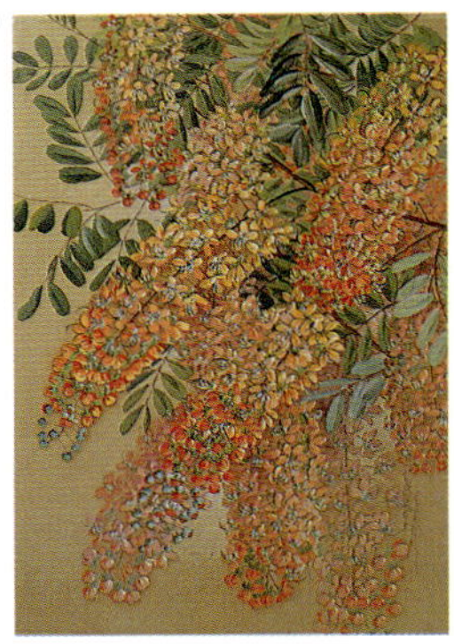

27

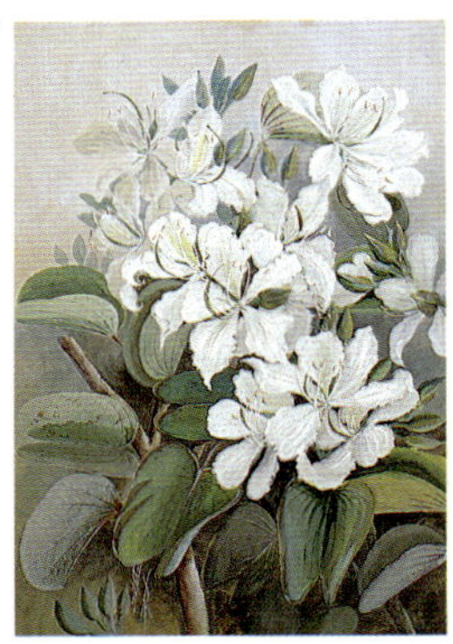

28

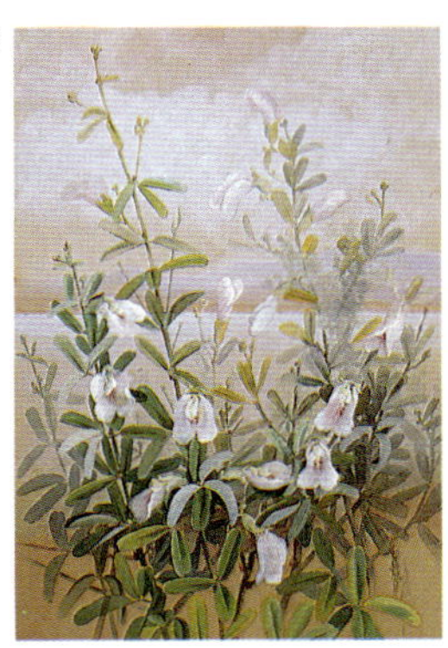

29

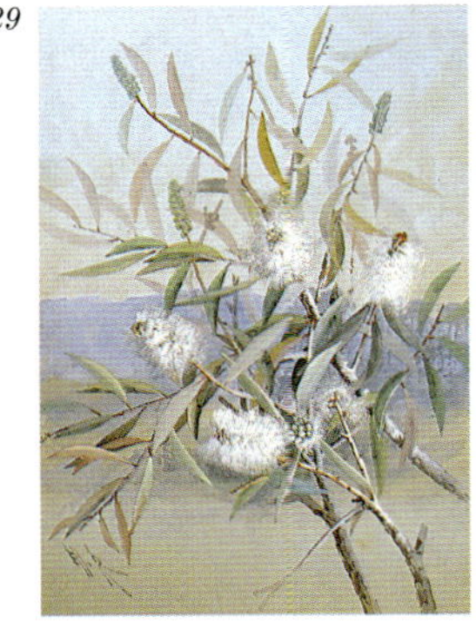

30

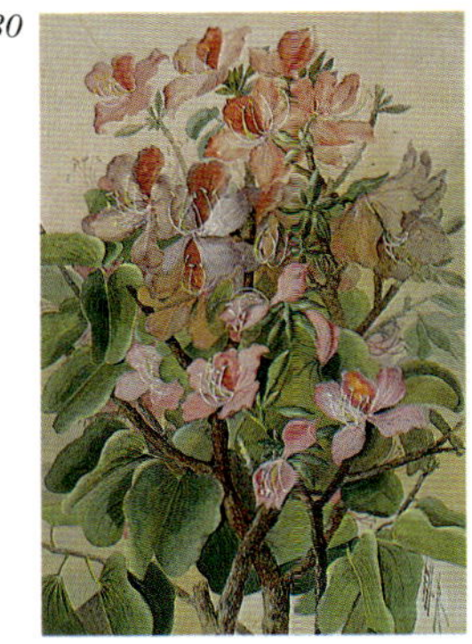

31

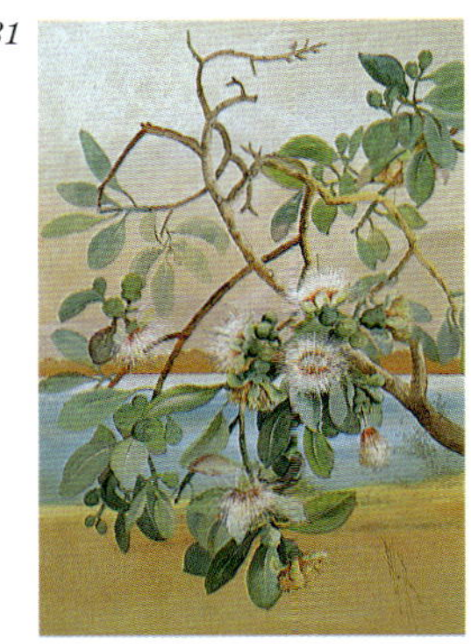

32

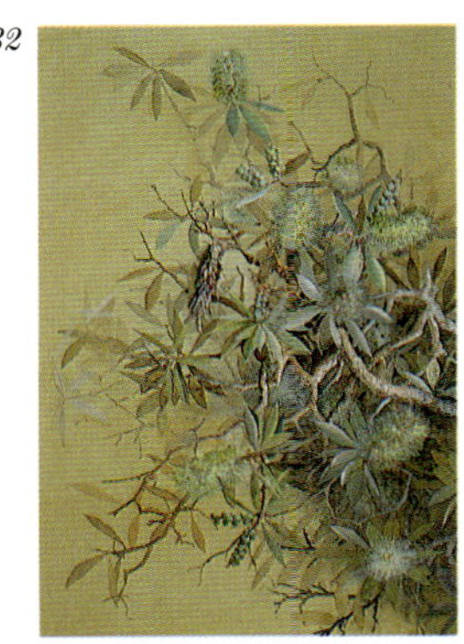

17
Jacksonia scoparia R.Br.
[Fabaceae]
DOGWOOD
E coast of Qld S of Mackay; NSW

Painted: (1911)

Numbered by artist reverse "55"
Titled by C.T. White reverse "Dogwood / Jacksonia / scoparia"

18
Twining vine - *Vigna vexillata* (L.) A.Rich. [Fabaceae]
WILD COW PEA
E Qld; NSW, NT; Africa

Support plant—*Themeda triandra* Forsskal [Poaceae]
KANGAROO GRASS
E Qld S from Cairns area; all Australian States

Painted: (1911)

Numbered by artist reverse "1"
Titled by C.T. White reverse "Vigna vexillata / the small pod Glycine sp. / the grass in the background / is Anthistiria ciliata / (Kangaroo Grass)"
Inscribed reverse "Brown and Crown Sts"

19
Castanospermum australe Hook. [Fabaceae]
BLACK BEAN; MORETON BAY CHESTNUT
E coast of Qld; NSW; widely cultivated

Painted: (1911)

Numbered by artist reverse "2"
Titled by C.T. White reverse "Castanospermum / Australe"

20
Austrosteenisia blackii (F.Muell.) Gees. [Fabaceae]
BLOOD VINE
Rainforests of Qld; NSW

Insect—*Anaphaeis java* Sparrman [Pieridae]
CAPER WHITE BUTTERFLY

Painted: (1911)

Numbered by artist reverse "46"
Titled by C.T. White reverse "Lonchocarpus / Blackii"

21
Crotalaria novae-hollandiae DC.
[Fabaceae]
NEW HOLLAND RATTLEPOD
Tropical Qld; NT, WA

Insect—*Hypolimnas bolina* L.
[Nymphalidae]
COMMON EGGFLY BUTTERFLY

Painted: (1911)

Numbered by artist reverse "53" (crossed out)
Titled by C.T. White reverse "Cassia / Mitchellii"

22
Cassia queenslandica C.T. White [Caesalpiniaceae]
QUEENSLAND CASSIA
Wet tropics of NE Qld

Painted: Cairns ranges (1891)

Inscribed by artist reverse "Queensland / Cairns ranges / 70"
Titled by F.v. Mueller reverse "Cassia Brewsteri, FvM"

Exhibited: Chicago 1893; Sydney 1910

Referred to: *Flower-Hunter*, p.51

23
Lysiphyllum cunninghamii (Benth.) de Wit [Caesalpiniaceae]
NATIVE BAUHINIA
Widespread in Qld except for C and S coast; NT, WA

Painted: (1911)

Numbered by artist reverse "50"
Titled by C.T. White reverse "Bauhinia / Cunninghamii / forma rosea"

24
Elaeagnus triflora Roxb.
[Elaeagnaceae]
MILLAA MILLAA
Rainforests of Qld N from Brisbane area

Insects—*Papilio demoleus* L.
[Papilionidae]
CHECKERED SWALLOWTAIL BUTTERFLY

Painted: (1911)

25
Petalostylis labicheoides R.Br.
[Caesalpiniaceae]
BUTTERFLY BUSH
C and W Qld; NSW, NT, WA, SA

Painted: (1911)

Numbered by artist reverse "53"
Titled by C.T. White reverse "Petalostyles / labicheoides"

26
Cassia queenslandica C.T. White
[Caesalpiniaceae]
QUEENSLAND CASSIA
Wet tropics of NE Qld

Painted: (1911)

Numbered by artist reverse "60"
Titled by C.T. White reverse "Cassia / Brewsteri"

27
Bauhinia variegata L.
[Caesalpiniaceae]
BUTTERFLY TREE
Introduced to Australia from E Asia and widely cultivated

Painted: (1911)

Numbered by artist reverse "26"
Titled by C.T. White reverse "Bauhinia / variegata var. candida"

28
Clitoria laurifolia Poiret
[Fabaceae]
E coast of Qld N from Townsville area

Painted: (1911)

29
Melaleuca quinquenervia (Cav.) S.T. Blake [Myrtaceae]
PAPER-BARKED TEA-TREE
Tea-tree swamps of E coast of Qld; NSW; New Guinea, New Caledonia

Painted: (1911)

Numbered by artist reverse "21"

30
Bauhinia purpurea L.
[Caesalpiniaceae]
BUTTERFLY TREE
Introduced to Australia and widely cultivated

Painted: (1911)

Numbered by artist reverse "30"
Titled by C.T. White reverse "Bauhinia / variegata"

31
Planchonia careya (F.Muell.) Knuth [Lecythidaceae]
COCKATOO APPLE; COCKY APPLE
C Qld and E coast N of Maryborough; NT, WA; New Guinea

Painted: probably Dungeness (1911)

Numbered by artist reverse "14"
Titled by C.T. White reverse "Careya / Australis"

32
Melaleuca quinquenervia (Cav.) S.T. Blake [Myrtaceae]
PAPER-BARKED TEA-TREE
Tea-tree swamps of E coast of Qld; NSW; New Guinea, New Caledonia

Insect—a larval case of a case-moth [Psychidae]

Painted: (1911)

Numbered by artist reverse "23"

33

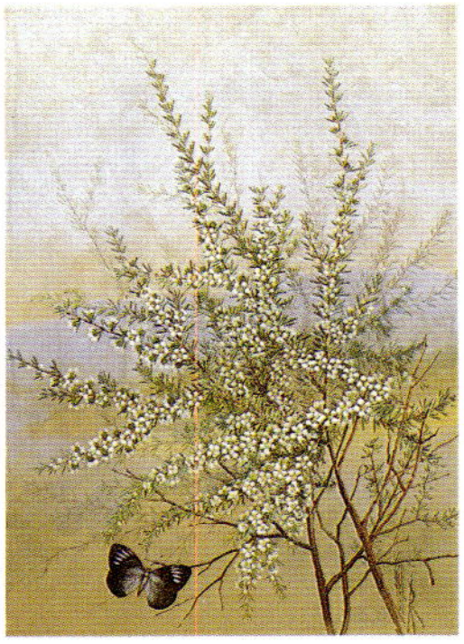

34

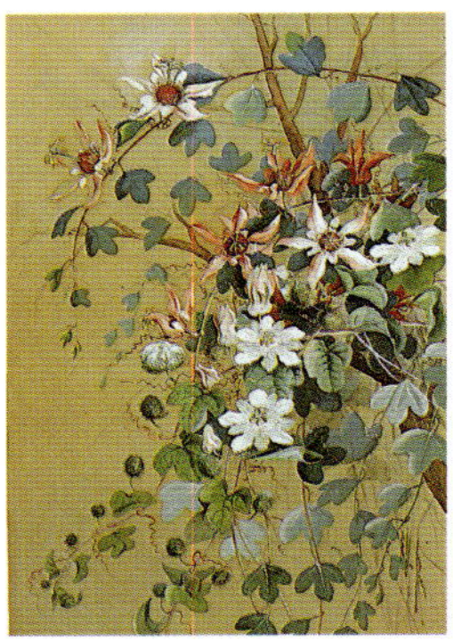

35

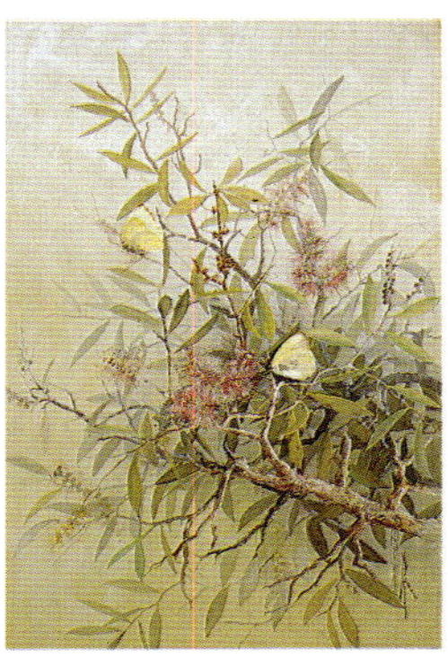

36

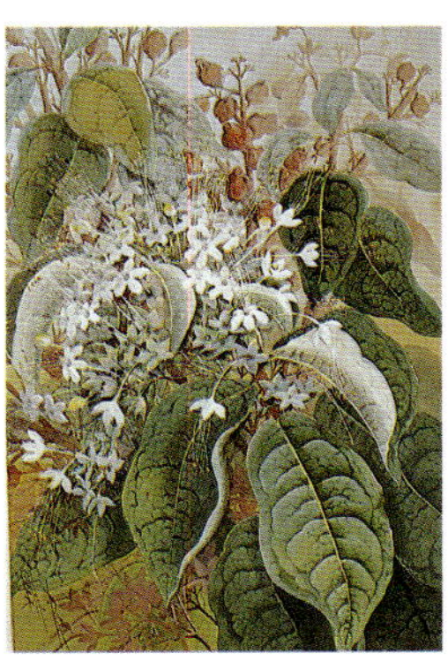

37

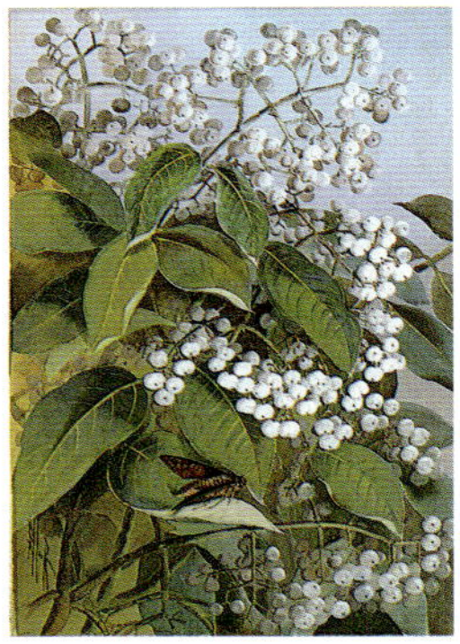

38

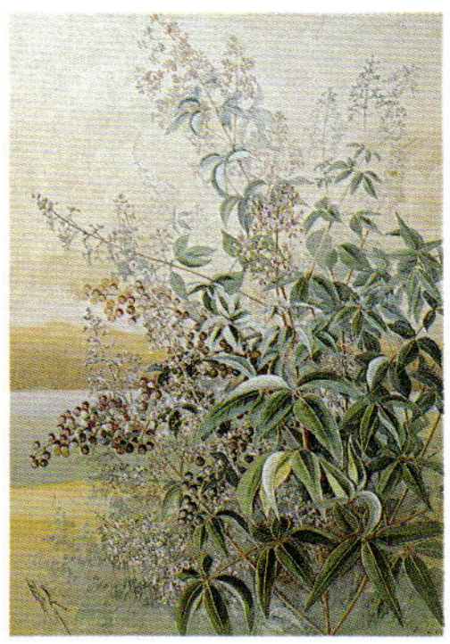

39

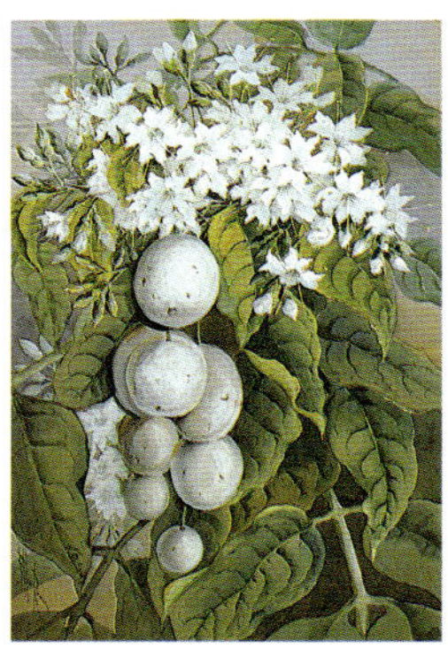

40

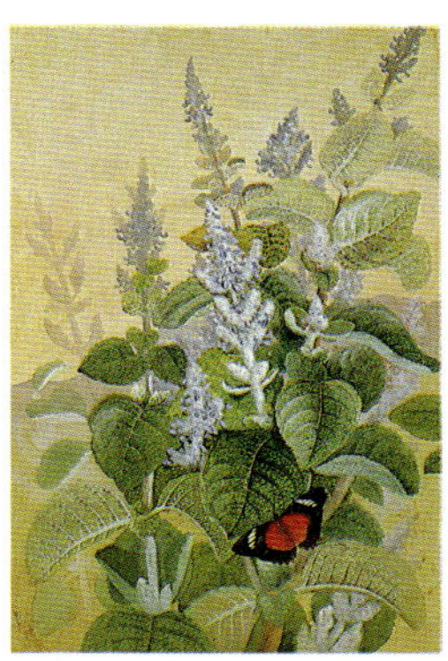

41

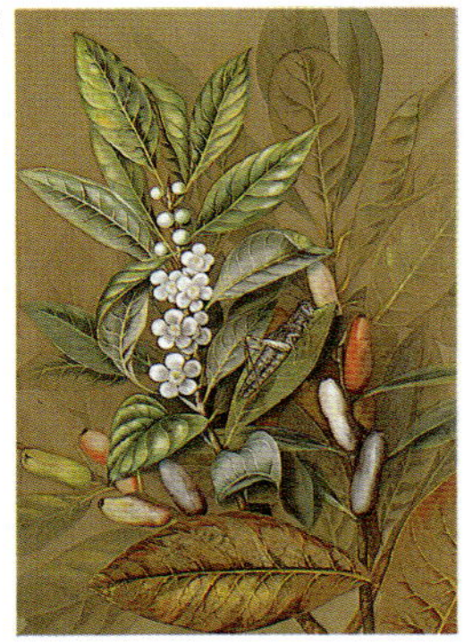

42

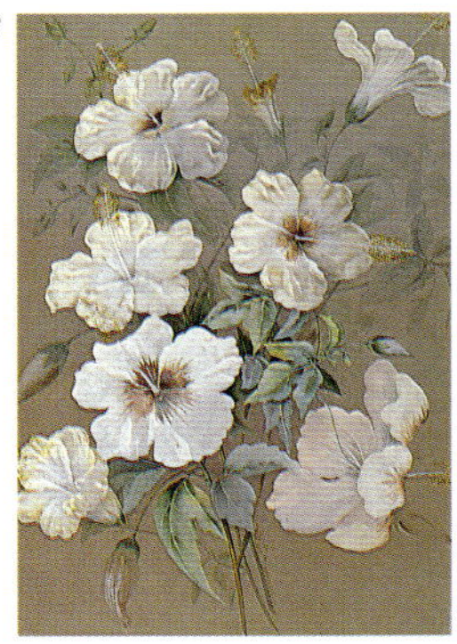

43

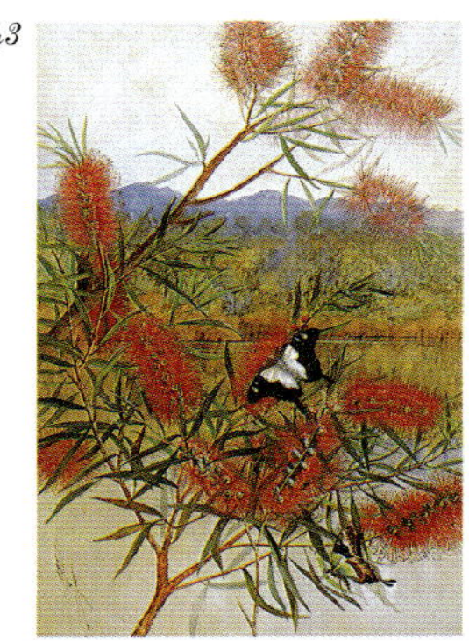

44

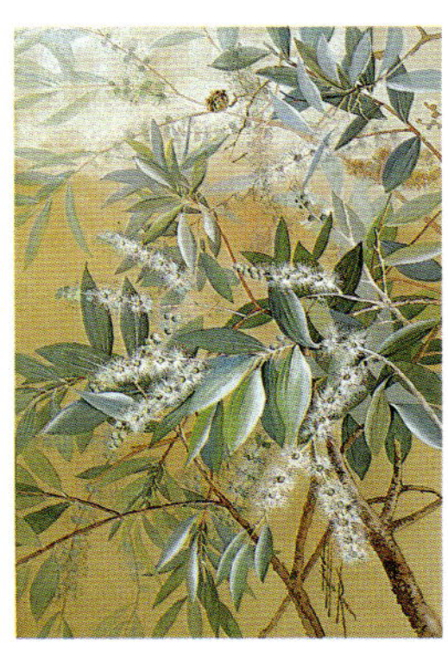

45

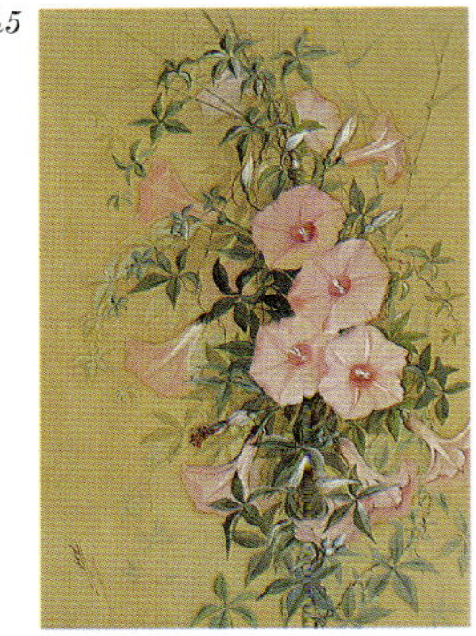

46

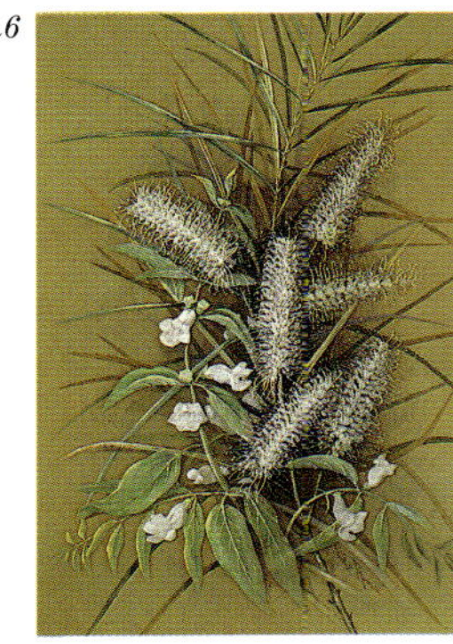

47

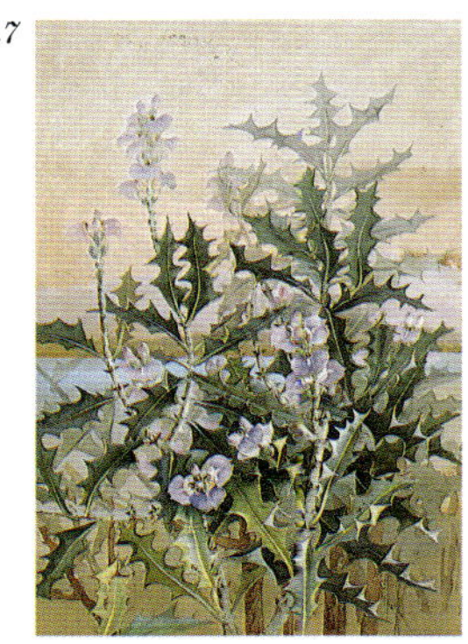

48

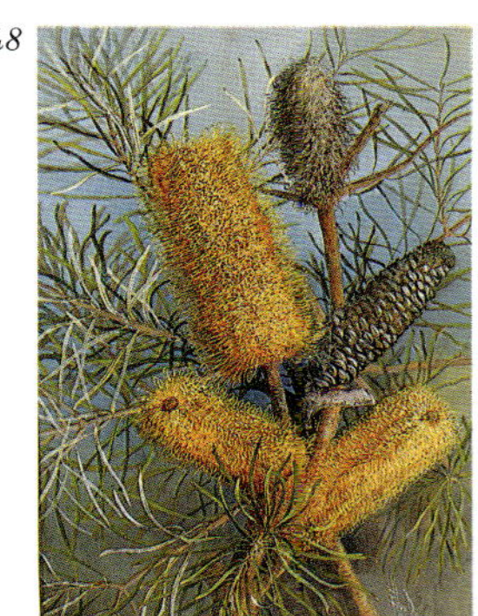

33
Leptospermum flavescens Smith [Myrtaceae]
WILD MAY
Widespread in E Qld; NSW, Vic., Tas.

Insect—*Delias* sp. [Pieridae]
A JEZABEL BUTTERFLY

Painted: (1911)

Numbered by artist reverse "28"
Titled by C.T. White reverse "Leptospermum / flavescens"

34
Top third and leaves at bottom right—*Passiflora aurantia* G. Forster [Passifloraceae]
ORANGE PASSION FLOWER
E coast of Qld N of Maryborough; NSW; New Guinea and other SW Pacific Is.

Centre and bottom left—*Passiflora foetida* L. [Passifloraceae]
STINKING PASSION FLOWER
Introduced to Australia and widely naturalised in Qld; NT, WA; West Indies, S America

Painted: (1911)

Numbered by artist reverse "39"
Titled by C.T. White reverse "Passiflora / aurantia / (pink)" and "Passiflora / foetida (white)"

35
Melaleuca quinquenervia (Cav.) S.T. Blake [Myrtaceae]
PAPER-BARKED TEA-TREE
Tea-tree swamps of E coast of Qld; NSW; New Guinea, New Caledonia

Insect—*Catopsilia* sp. [Pieridae]
A MIGRANT BUTTERFLY

Painted: (1911)

Numbered by artist reverse "22"

36
Clerodendrum cunninghamii Benth. [Verbenaceae]
NATIVE CLERODENDRUM
Coastal Qld N of Maryborough; NT

Painted: (1911)

Inscribed by artist reverse "56 / Clerodendron / Cunninghamia" (sic.)

Similar work reproduced: Bailey, plate 11

37
Psychotria coelosperma F.M. Bailey [Rubiaceae]
E coast of Qld N of Ingham

Insect—*Gnathothlibus erotus* (Cramer) [Sphingidae]
A HAWK MOTH

Painted: (1911)

Numbered by artist reverse "31"
Titled by C.T. White reverse "Psychotria / coelospermum"

38
Vitex trifolia L. [Verbenaceae]
E coast of Qld; NSW; Africa, India, SE Asia, Japan, Pacific Is.

Painted: (1911)

Numbered by artist reverse "63"
Titled by C.T. White reverse "Vitex / trifolia"

39
Faradaya splendida F.Muell. [Verbenaceae]
Rainforests of Qld N of Townsville

Painted: (1911)

Inscribed by artist reverse "51 / Faradaya / splendida"

40
Plectranthus graveolens R.Br. [Lamiaceae]
NATIVE COLEUS
E Qld; NSW

Insect—*Cethosia cydippe* L. [Nymphalidae]
RED LACEWING BUTTERFLY

Painted: (1911)

Numbered by artist reverse "49"
Titled by C.T. White reverse "Plectranthus / parviflorus / var. major"

41
Rhodomyrtus macrocarpa Benth. [Myrtaceae]
FINGER CHERRY
Rainforests of Qld N of Ingham

Insect—possibly *Valanga irregularis* (Walker) or an *Adreppus* sp. [Acrididae]
A GRASSHOPPER

Painted: Somerset (1891-92)

Inscribed by artist reverse "Somerset No.4 / Rhodomyrtus macrocarpa"
Inscribed reverse "Rumau-ju"

Exhibited: Possibly Chicago 1893; London 1896

42
Hibiscus rosa-sinensis L. [Malvaceae]
HIBISCUS
Introduced to Australia and widely cultivated; world-wide

Painted: New Guinea (1892)

Inscribed by artist reverse "New Guinea No.5 / Hibiscus Rosa / 82"
Numbered lower left corner "123"

Exhibited: London 1896; Sydney 1910

43
Callistemon viminalis (Gaertner) G.Don ex Loudon [Myrtaceae]
WEEPING BOTTLEBRUSH
Coastal Qld; NSW

Insects—*Graphium macleayanum* (Leach) [Papilionidae]
MACLEAY'S SWALLOWTAIL BUTTERFLY

Painted: (1911)

Numbered by artist reverse "59"
Titled by C.T. White reverse "Callistemon / Lanceolatus"

44
Melaleuca dealbata S.T. Blake [Myrtaceae]
WEEPING TEA-TREE
Tea-tree swamps of E coast of Qld N of Maryborough; NT, WA

Painted: (1911)

Numbered by artist reverse "49" and "54"
Titled by C.T. White reverse "Melaleuca / leucadendron"

45
Ipomoea cairica (L.) Sweet [Convolvulaceae]
COAST MORNING GLORY
E coast of Qld S of Cairns; NSW; tropical Asia, Africa

Painted: (1911)

Numbered by artist reverse "57"
Titled by C.T. White reverse "Ipomoea / palmata"

46
Foreground—*Josephinia imperatricis* Vent. [Pedaliaceae]
Cape York Peninsula in Qld; NT, WA; New Guinea

Background—*Grevillea* sp. [Proteaceae]

Painted: Somerset (1891-92)

Inscribed by artist reverse "Somerset / No. 6"
Titled by F.v. Mueller reverse "1. Josephinia grandiflora, R. Brown / 2. Grevillea polystachya, R. Brown"

Exhibited: Chicago 1893

47
Acanthus ilicifolius L. [Acanthaceae]
Mangrove forests of coastal Qld; NT

Painted: (1911)

Numbered by artist reverse "44"
Titled by C.T. White reverse "Acanthus / ilicifolius"

48
Banksia spinulosa var. *collina* (R.Br.) A.S. George [Proteaceae]
GOLDEN CANDLESTICKS
E coast of Qld; NSW, Vic.

Painted: (1911)

Numbered by artist reverse "25"
Titled by C.T. White reverse "Banksia / collina"

Reproduced: *Christmas Queenslander*, 7 December 1918, colour supplement

49

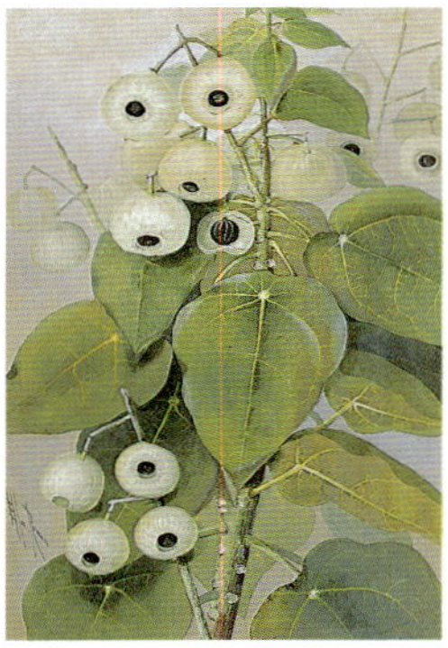

50

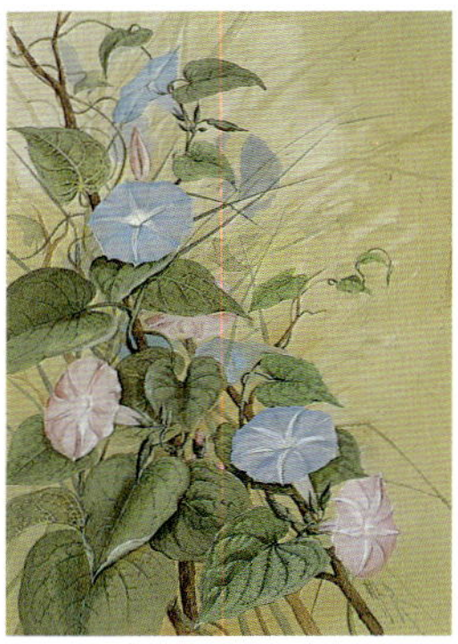

51

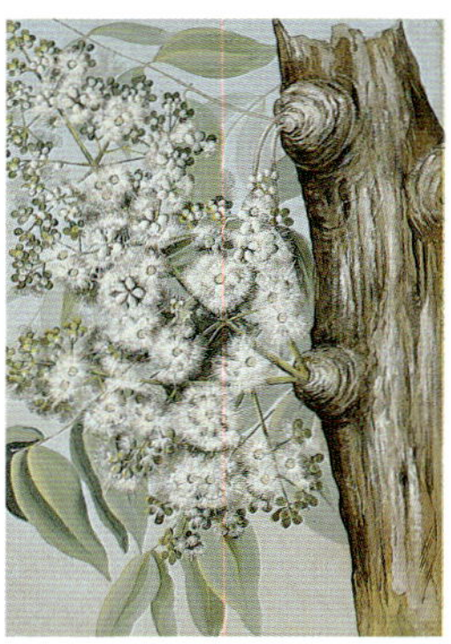

52

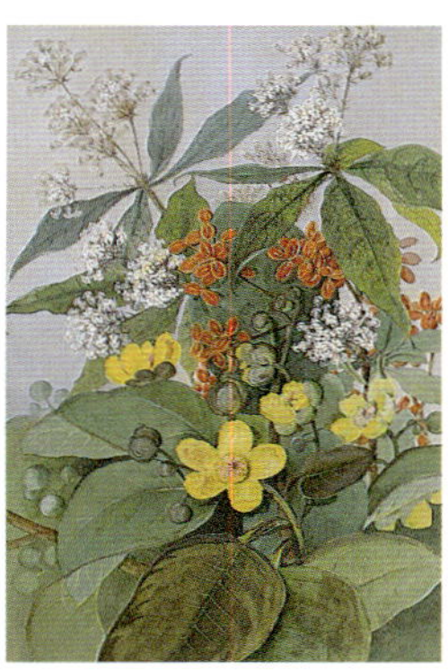

53

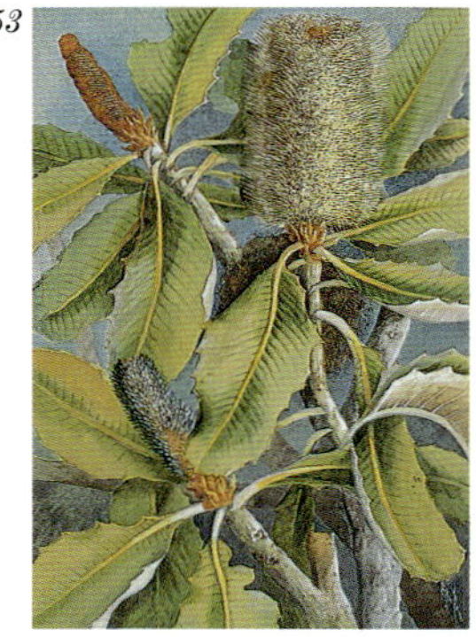

54

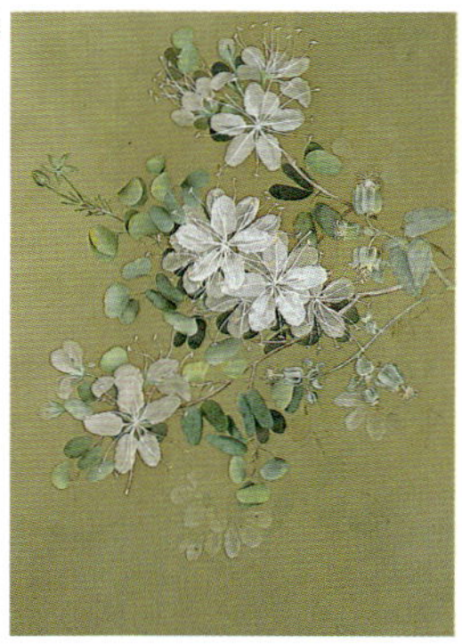

55

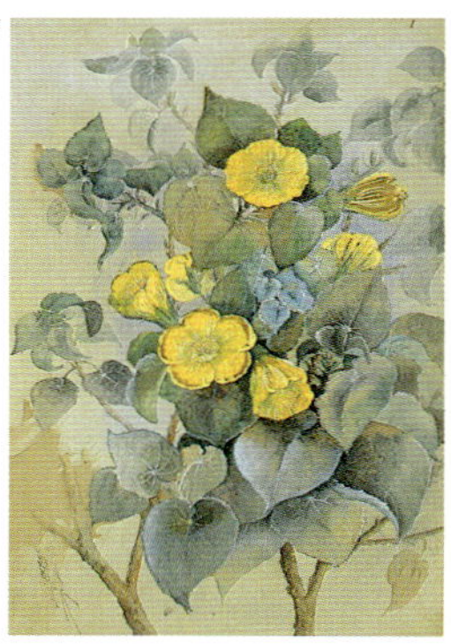

56

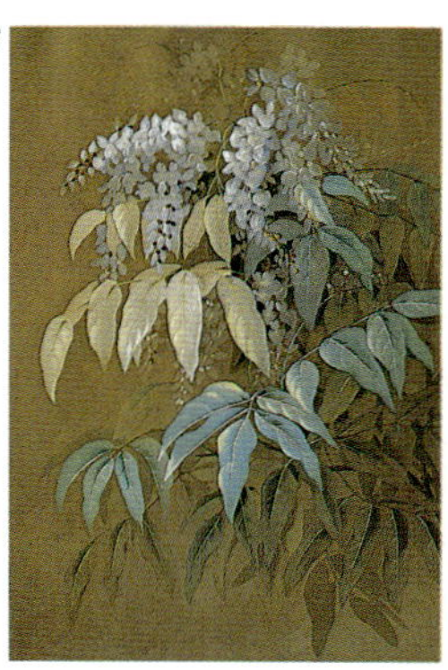

57

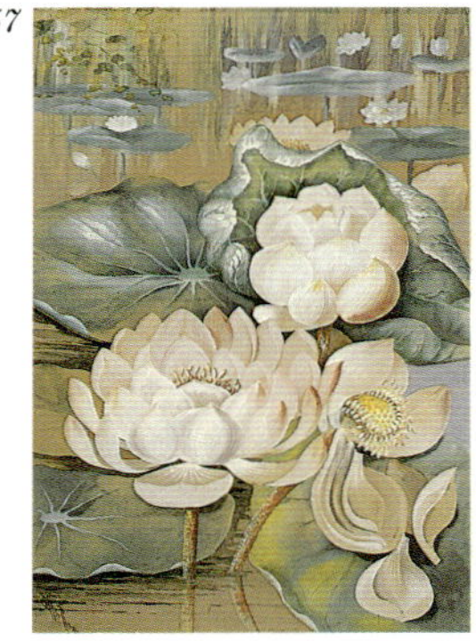

58

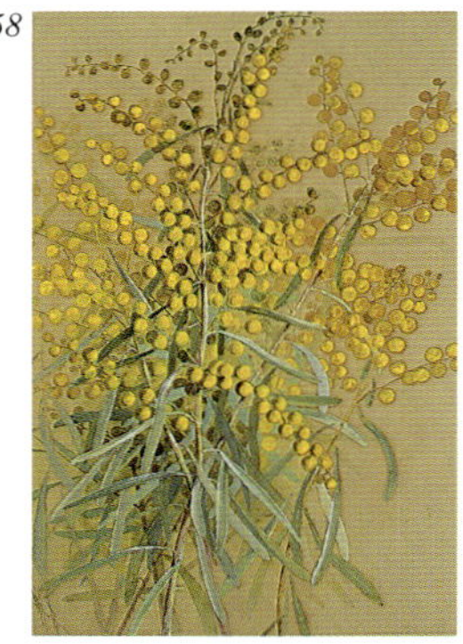

59

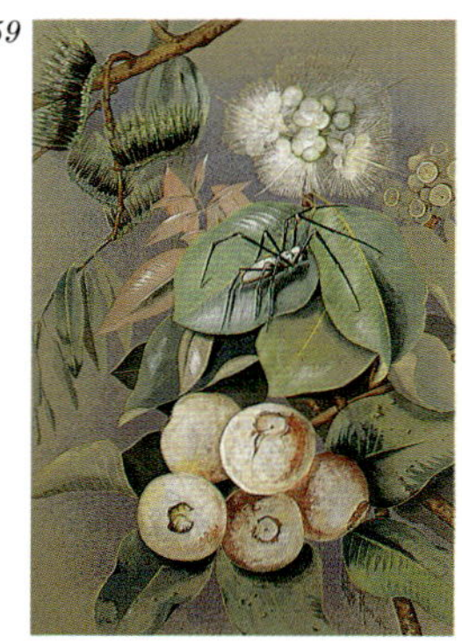

60

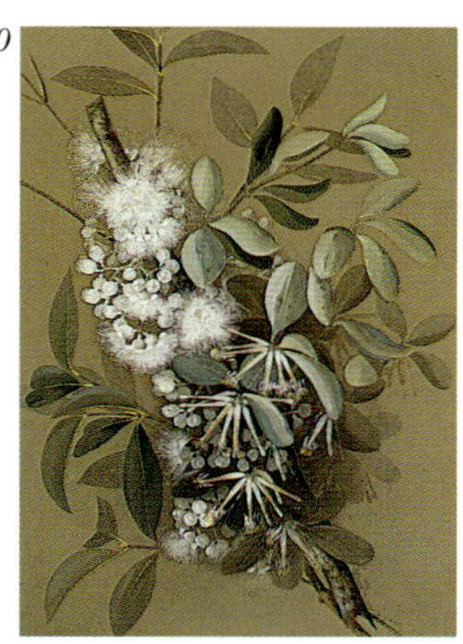

61

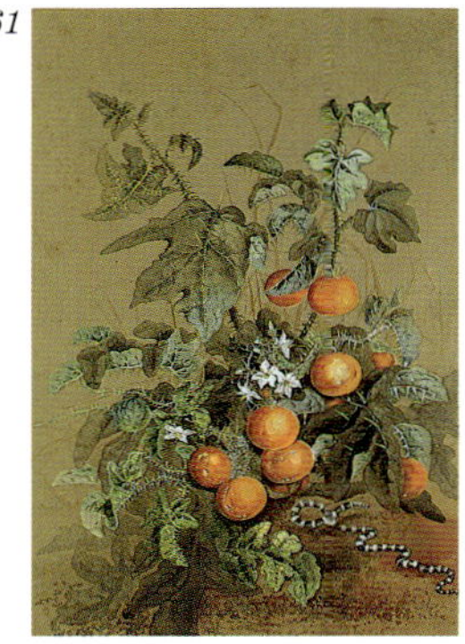

62

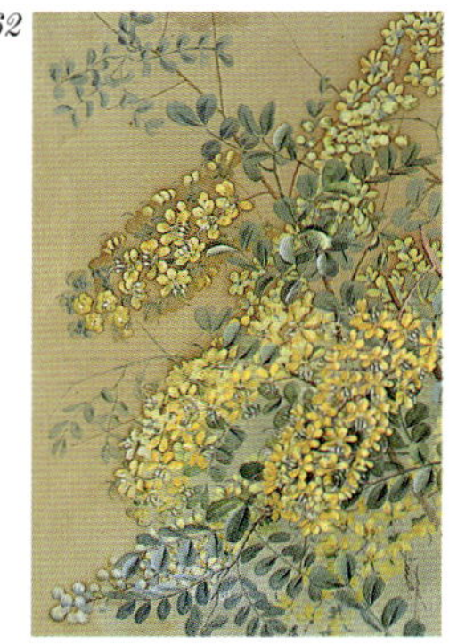

63

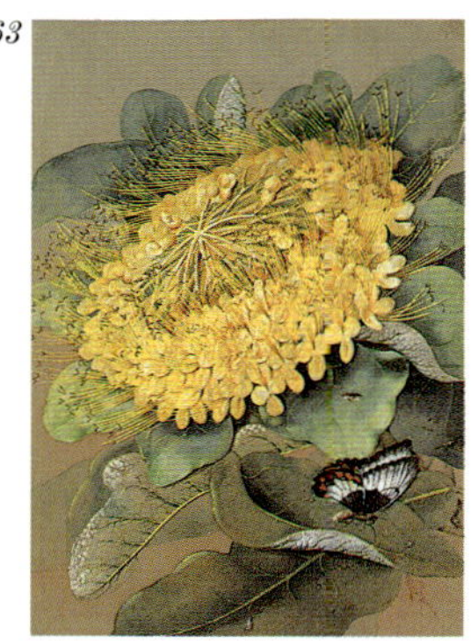

64

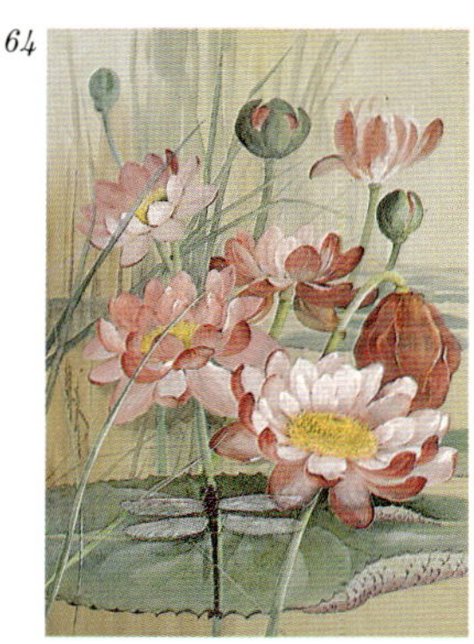

49
Hernandia peltata Meisner [Hernandiaceae]
Strand forests of Qld N of Townsville; Pacific Is., SE Asia, E Africa

Painted: (1911)

Numbered by artist reverse "19"

Similar work reproduced: Bailey, plate 13; Tardent p. 49; *Muses Magazine*, April 1928

50
Ipomoea indica (Burman) Merr. [Convolvulaceae]
BLUE MORNING GLORY
E coast of Qld; NSW, NT; Pantropical

Painted: (1911)

Numbered by artist reverse "55"
Titled by C.T. White reverse "Ipomoea / rubro-caerulea"

51
Syzygium branderhorstii Lauterb. [Myrtaceae]
BUMPY SATIN ASH
Rainforests of Qld N of Townsville

Painted: Somerset (1891-92)

Inscribed by artist reverse "Somerset / N. Australia / 81"

Exhibited: Sydney 1910

52
Foreground—*Dillenia alata* (R.Br. ex DC.) Martelli [Dilleniaceae]
GOLDEN GUINEA TREE
Rainforests of Qld N of Townsville; NT

Background - *Pittosporum rubiginosum* Cunn. [Pittosporaceae]
Rainforests of Qld N of Townsville

Painted: (1911)

Numbered by artist reverse "4"
Titled by C.T. White reverse "Pittosporum / rubiginosum" and "Wormia alata"

Similar work reproduced: Bailey, plate 2; Tardent, p. 13; *Muses Magazine*, August and October-November 1928

53
Banksia robur Cav. [Proteaceae]
BROAD-LEAVED BANKSIA
E coast of Qld; NSW

Painted: (1911)

Numbered by artist reverse "40"
Titled by C.T. White reverse "Banksia / latifolia"

54
Foreground—*Abutilon micropetalum* Benth. [Malvaceae]
E coast of Qld N of Bundaberg

Background—*Lysiphyllum hookeri* (F.Muell.) Pedley [Caesalpiniaceae]
PEGUNNY
Coastal Qld N of Bundaberg; NSW; Asia

Top right hand side—?

Painted: (1891-92)

Numbered by artist centre right "2"
Titled by F.v. Mueller reverse "1 Bauhinia Hookeri F.v.M. / 2 Abutilon micropetalum, Bentham"

Exhibited: London 1896; Sydney 1910

55
Abutilon indicum (L.) Sweet [Malvaceae]
LANTERN BUSH
Coastal tropical Qld

Painted: (1911)

Numbered by artist reverse "33"

56
Millettia megasperma (F.Muell.) Benth. [Fabaceae]
NATIVE WISTARIA
Rainforests of Qld S of Bundaberg; NSW

Painted: (1891-92)

Inscribed by artist reverse "76 / Queensland"

Exhibited: Sydney 1910

57
Nelumbo nucifera Gaertner [Nelumbonaceae]
PINK LOTUS-LILY
Swamps and waterways of N Qld; NT

Painted: after the original, Herbert River 1887

Titled by artist reverse "Nelumbo nucifera"
Numbered lower left corner "216"

Exhibited: London 1896; Melbourne 1904; Sydney 1910

58
Acacia sp. [Mimosaceae]

Painted: (1891-92)

Numbered by artist reverse "100"
Titled reverse "Variety Wattle"

Exhibited: Sydney 1910

59
Top—*Decaisnina brittenii* (Blakely) Barlow [Loranthaceae]
A MISTLETOE
Coastal Qld N of Cairns; NT

Bottom—*Syzygium suborbiculare* (Benth.) T. Hartley & Perry [Myrtaceae]
ROLY-POLY SATIN ASH
Coastal Qld N of Cairns; NT, WA

Spider—*Nephila* sp.? *maculata* [Araneidae]
GOLDEN ORB-WEB SPIDER

Painted: Somerset (1891)

Inscribed by artist reverse "Somerset / No 10 / The natives chew the leaves of / Eugenia when bitten by brown snakes / (illeg.)" and "Eugenia macrocarpa, Roxburgh"

Exhibited: Melbourne 1904

Referred to: *Flower-Hunter*, p.142

60
Upper spray flowers—*Syzygium cormiflorum* (F.Muell.) B. Hyland [Myrtaceae]
BUMPY SATIN ASH
Rainforests of Cape York Peninsula in Qld

Lower tubular flowers—*Amyema congener* (Sieber ex Schultes & J.H. Schultes) Tieghem [Loranthaceae]
A MISTLETOE
E coast of Qld; NSW

Painted: Johnstone River 1887

Inscribed reverse "Q51 Johnstone River" and "Queensland / Eugenia"

Exhibited: Sydney 1910

61
Solanum capsicoides All. [Solanaceae]
DEVIL'S APPLE
Introduced to Qld from C America and widely naturalised; NSW

Snake—*Vermicella annulata* (Gray) [Elapidae]
BANDY-BANDY SNAKE

Painted: Mackay 1887

Inscribed by F.v. Mueller reverse "Solanum ciliatum Lamarck / Immigrated / from South / America"
Inscribed reverse "Q24 Mackay"

Exhibited: Sydney 1910

62
Cassia queenslandica C.T. White [Caesalpiniaceae]
QUEENSLAND CASSIA
Wet tropics of NE Qld

Painted: (1911)

Numbered by artist reverse "37"
Titled by C.T. White reverse "Cassia Brewsteri / var. sylvestris"

Similar work reproduced: Bailey, plate 6

63
Deplanchea tetraphylla (R.Br.) F.Muell. [Bignoniaceae]
GOLDEN BOUQUET TREE
E coast of Qld N of Townsville

Insect—*Papilio aegeus* Donovan [Papilionidae]
ORCHARD SWALLOWTAIL BUTTERFLY

Painted: Somerset (1891-92)

Inscribed by artist reverse "Somerset, Cape York / 74"
Titled by C.T. White reverse "Diplanthera"

Exhibited: Sydney 1910

64
Nymphaea gigantea J.D. Hook. [Nymphaeaceae]
GIANT WATERLILY
Swamps and waterways of mainly coastal Qld; NSW, NT

Insect—*Petalura ingentissima* Tillyard [Petaluridae]
A DRAGONFLY

Painted: (1911)

Numbered by artist reverse "64"

65
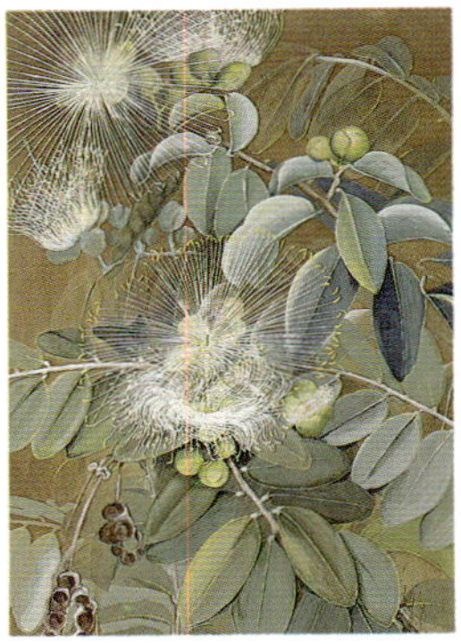

66
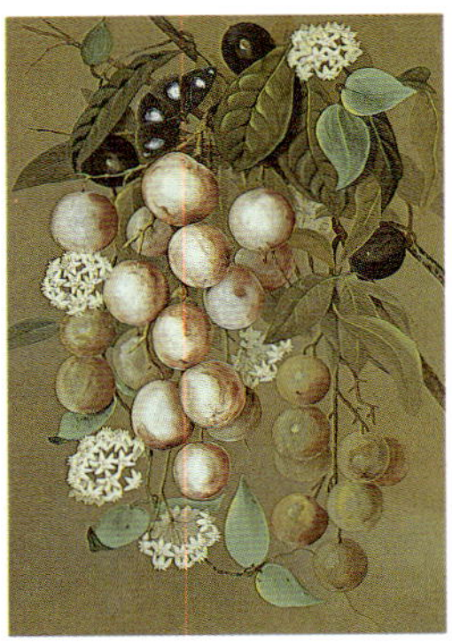

67
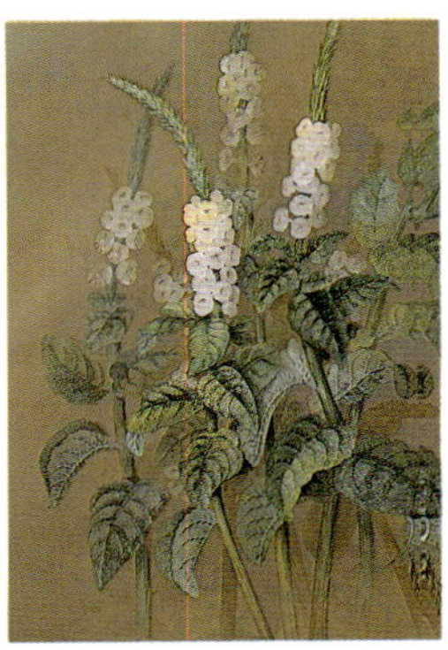

68
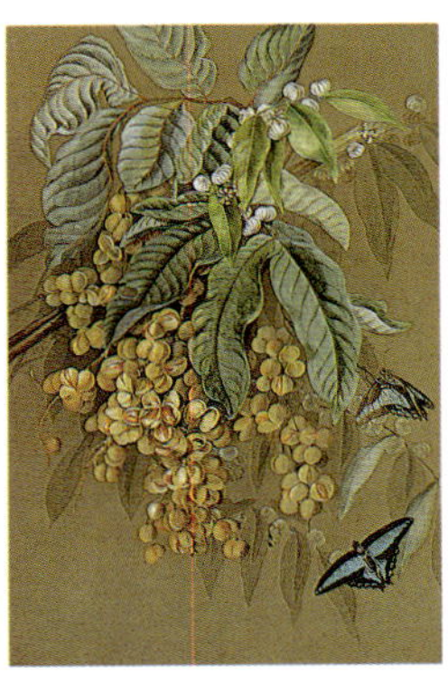

69
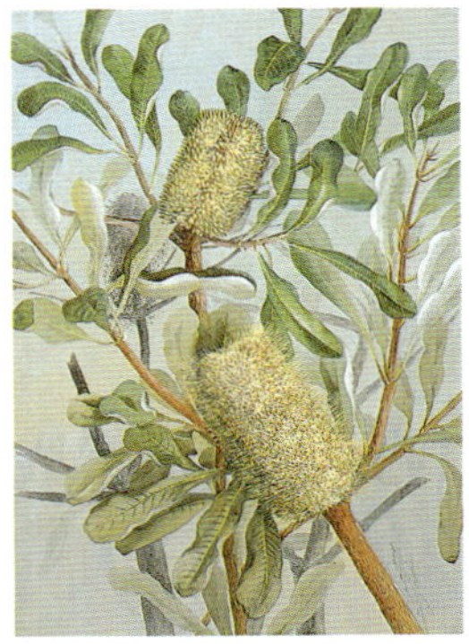

70
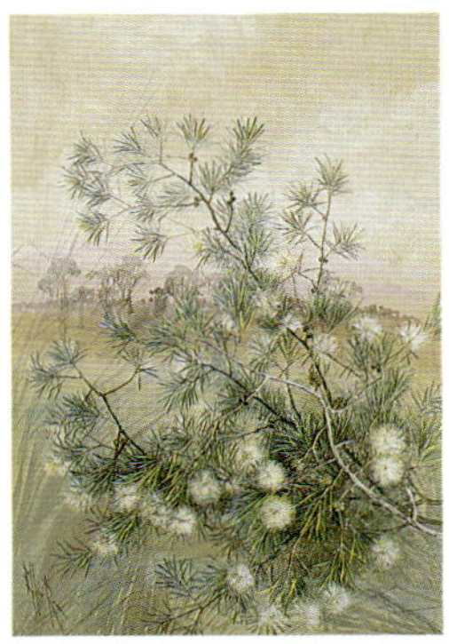

71
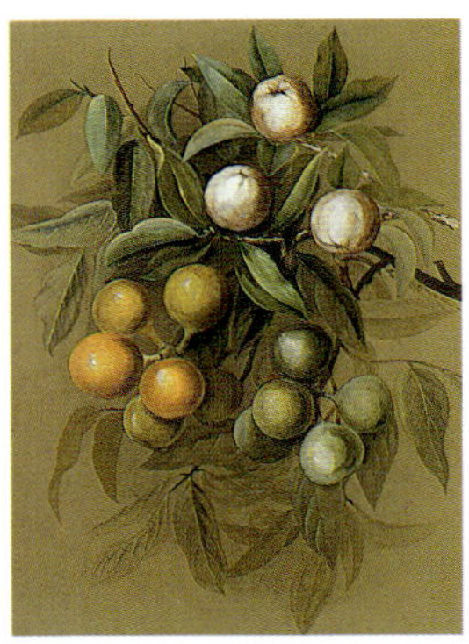

72
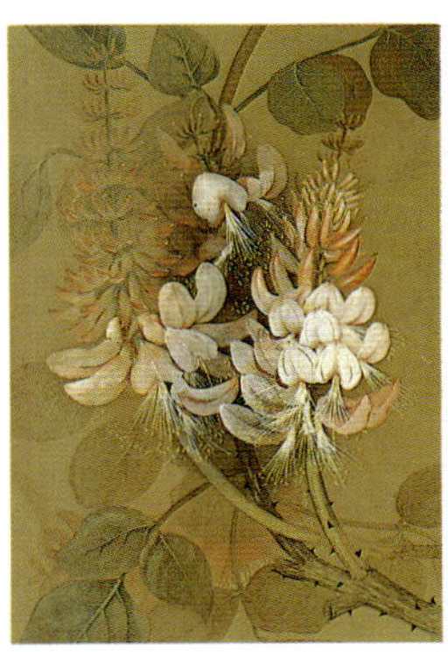

73
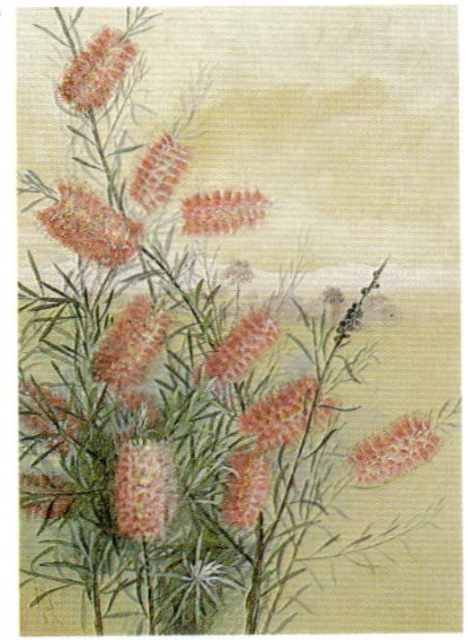

74
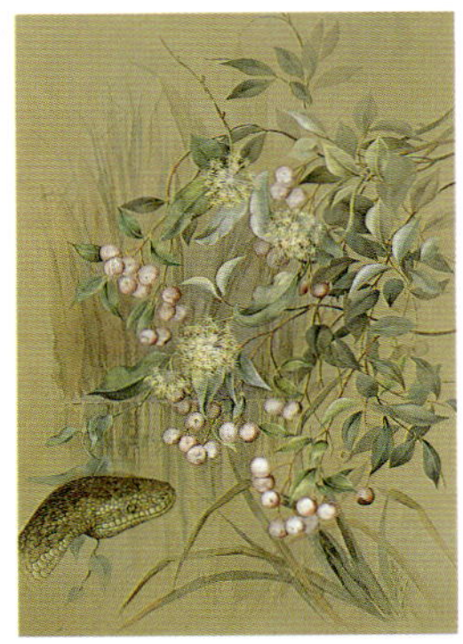

75
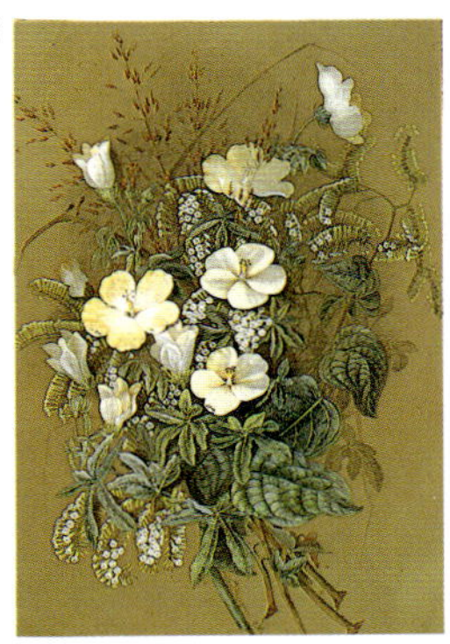

76
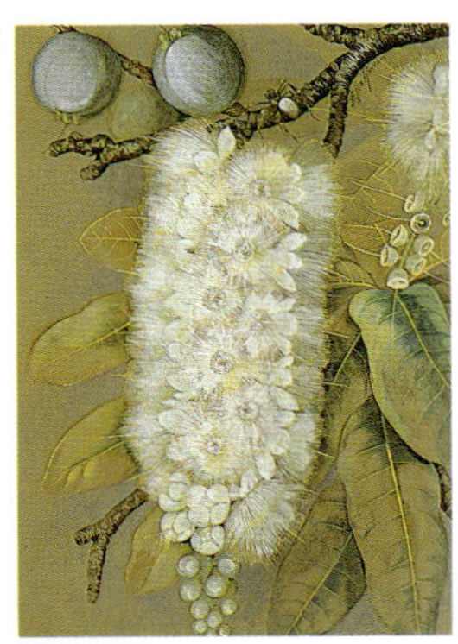

77
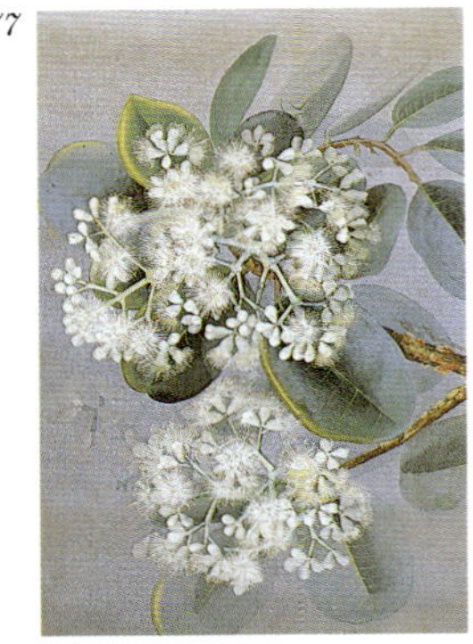

78
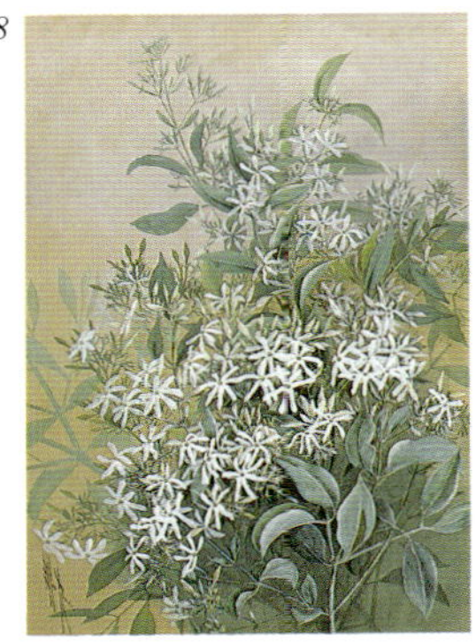

79
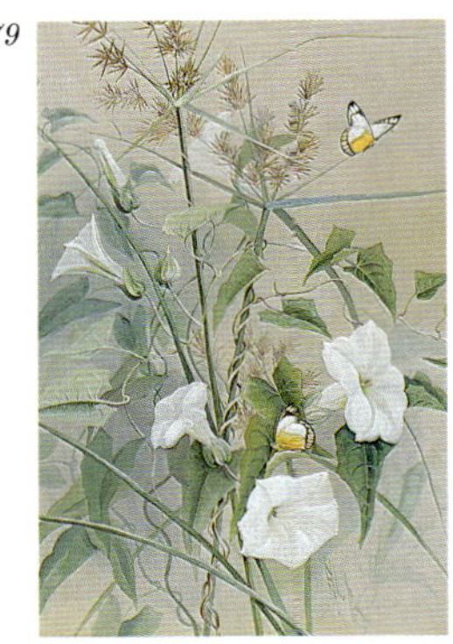

80
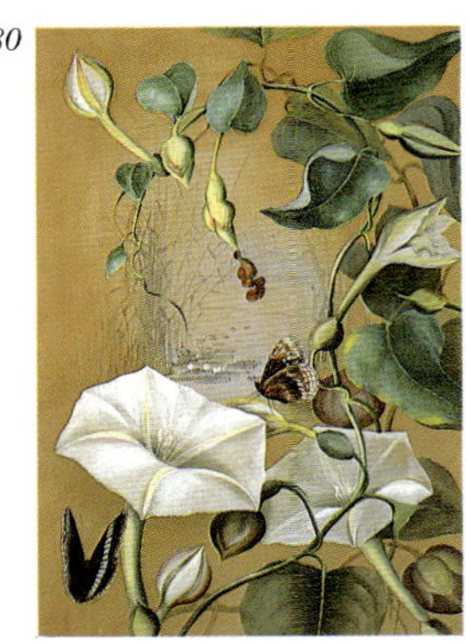

65
Capparis flexuosa (L.) L. [Capparaceae]
LIMBER CAPER
C America and West Indies

Painted: West Indies (c.1901)

Inscribed by artist reverse "68 / Capparis cynophallophora / Antigua"

66
Flowers—*Hoya nicholsoniae* F.Muell. [Asclepiadaceae]
NICHOLSON'S WAX VINE
E coast of Qld N from Townsville area

Light fruits—*Syzygium sayeri* (F.Muell.) B. Hyland [Myrtaceae]
PINK SATIN ASH
Rainforests of Cape York Peninsula in Qld

Dark fruits—*Niemeyera prunifera* (F.Muell.) F. Muell. [Sapotaceae]
PLUM BOXWOOD
E coast of Qld N from Townsville area

Insect—*Hypolimnas bolina* L. [Nymphalidae]
COMMON EGGFLY BUTTERFLY

Painted: Johnstone River 1887

Numbered by artist reverse "73"
Titled by F.v. Mueller reverse "1 Hoya nicholsoniae / 2 Eugenia sp"
Inscribed reverse "Q63 Johnstone River" and "Top"

Exhibited: Probably Melbourne 1888; Sydney 1910

67
Stachytarpheta mutabilis (Jacq.) Vahl [Verbenaceae]
PINK SNAKEWEED
Introduced to Qld and naturalised along coast N of Maryborough, cultivated

Painted: Cooktown (1891-92)

Inscribed by artist reverse "Cooktown No. 1 / 69 / Stachytarpheta Mutabilis"

Exhibited: London 1896; Melbourne 1904; Sydney 1910

68
Smaller leaves and entire fruits—*Glochidion ferdinandii* (Muell. Arg.) F.M. Bailey [Euphorbiaceae]
CHEESE TREE
E coast of Qld; NSW, NT, WA

Compound leaves and opened fruits—*Diploglottis diphyllostegia* (F.Muell.) F.M. Bailey [Sapindaceae]
E coast of Qld

Insects—*Graphium sarpedon* L. [Papilionidae]
BLUE TRIANGLE BUTTERFLY

Painted: Macknade Plantation, Herbert River 1887

Titled by F.v. Mueller reverse "1 Phyllanthus Ferdinandi F. Mueller / 2 Harpullia"
Inscribed reverse "Q37 Macknade (Herbert River)"
Numbered lower left corner "142"

Exhibited: Sydney 1910

69
Banksia integrifolia L.f. [Proteaceae]
COASTAL BANKSIA
E coast of Qld; NSW, Vic., Tas.

Painted: (1911)

Numbered by artist reverse "3"
Titled by C.T. White reverse "Banksia / integrifolia"

70
Melaleuca nodosa (Gaertner) Smith [Myrtaceae]
PRICKLYLEAF PAPERBARK
E coast of Qld S of Mackay; NSW, SA

Painted: (1891-92)

Numbered lower left corner "332"

Exhibited: Sydney 1910

71
Upper twig with three single fruits—? *Syzygium* sp. [Myrtaceae]

Lower twig with fruit clusters left and below—*Ficus racemosa* L. [Moraceae]
CLUSTER FIG
E coast of Qld N of Maryborough

Painted: Johnstone River 1887

Numbered by artist reverse "78"
Titled by F.v.Mueller reverse "1 Ficus glomerata, Willdenow / 2"
Inscribed reverse "Q57 Johnstone River"

Exhibited: Sydney 1910

72
Erythrina variegata L. [Fabaceae]
A CORAL TREE
E coast of Qld; New Guinea and other W Pacific Is., Asia, India, Philippines, Indonesia

Painted: Jervis Is. (1891)

Inscribed by artist reverse "Jervis Island No. 9" (illeg.)
Titled by F.v. Mueller reverse "Erythrina"

Exhibited: Possibly London 1896; Melbourne 1904

73
Callistemon viminalis (Gaertner) G.Don ex Loudon [Myrtaceae]
WEEPING BOTTLEBRUSH
Coastal Qld; NSW

Painted: (1911)

74
Flowers on vine - *Smilax australis* R.Br. [Smilacaceae]
BARBWIRE VINE
E Qld; NSW, NT, Vic.

Fruits on twig—*Acmena smithii* (Poiret) Merr. & Perry [Myrtaceae]
LILLY PILLY
E Qld; NSW, Vic.

Snake—*Python spilotes* (Lacépède) [Boidae]
CARPET SNAKE

Painted: Normanby Station, Harrisville (1891)

Inscribed by artist reverse "Normanby / Queensland / 71"
Inscribed reverse "Eugenia berries" and "Top"
Numbered lower left corner "126"

Exhibited: Sydney 1910

75
Large flowers and pinnatifid leaves—*Abelmoschus moschatus* subsp. *tuberosus* (Span.) Borssum Waalkes [Malvaceae]
E coast of Qld; NT, WA

Small flowers and entire leaves—*Argusia argentea* (L.f.) I.M. Johnson [Boraginaceae]
E coast of Qld N from Bundaberg area; Asia, SW Pacific Is.

Grass—*Sorghum plumosum* (R.Br.) Beauv. [Poaceae]
PLUME SORGHUM
All Qld N from Rockhampton area; NT

Painted: Mackay 1887

Numbered by artist reverse "85"
Titled by F.v. Mueller reverse "1, Andropogon australis, Sprengel / 2, Hibiscus rhodopetalus, F.v.Mueller / 3, Tournefortia sarmentosa, Lamarck"
Inscribed reverse "Q17 Mackay"
Numbered lower left corner "460"

Exhibited: Sydney 1910

76
Barringtonia calyptrata (Miers) F.M. Bailey [Lecythidaceae]
Coastal Qld N of Townsville; New Guinea

Spider—*Nephila* sp. ?*plumipes* [Araneidae]
A GOLDEN ORB-WEB SPIDER

Painted: Thursday Is. (1891–92)

Inscribed by artist reverse "Thursday Island / 72 / Eugenia" (Eugenia later crossed out)
Titled reverse "Barringtonia / racemosa"

77
Syzygium forte (F.Muell.) B. Hyland subsp. *forte* [Myrtaceae]
FLAKY-BARKED SATIN ASH (timber); WHITE APPLE (plant)
E coast of Qld N of Ingham; NT

Painted: Somerset (1891-92)

Inscribed by artist reverse ?"43 / Somerset" (almost illeg.)
Numbered lower right corner "252"

Exhibited: Sydney 1910

78
Tabernaemontana pandacaqui Lam. [Apocynaceae]
E coast of Qld N from Bundaberg area; NT; New Guinea and other SW Pacific Is., SE Asia

Painted: (1911)

Numbered by artist reverse "6"
Titled by C.T. White reverse "Tabernaemontana / orientalis"

79
Twiner—*Calystegia sepium* (L.) R.Br. [Convolvulaceae]
NSW, Vic., Tas.; New Zealand, Europe, Africa, America

Support plant—*Cyperus lucidus* R.Br. [Cyperaceae]
A SEDGE
E coast of Qld; NSW, NT, Vic., Tas.; Pantropical

Insects—*Delias mysis* (Fabricius) [Pieridae]
UNION JACK BUTTERFLY

Painted: Vic. (c.1910)

Inscribed by artist reverse "79 / Ipomea / Vic" (sic.)

Exhibited: Sydney 1910

80
Ipomoea macrantha Roemer & Schultes [Convolvulaceae]
E coast of Qld N of Maryborough; NSW, NT, WA

Insects—*Hypolimnas alimena* (L.) [Nymphalidae]
BLUE-BANDED EGGFLY BUTTERFLY

Painted: Bloomfield River (1892)

Inscribed by artist reverse "Ipomea Grandiflora / Bloomfield River / Queensland' (sic.)

Exhibited: Possibly Chicago 1893; London 1896; Melbourne 1904; Sydney 1910

81

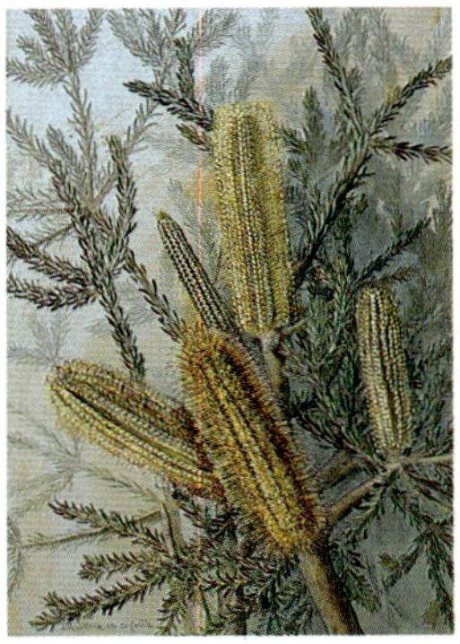

82

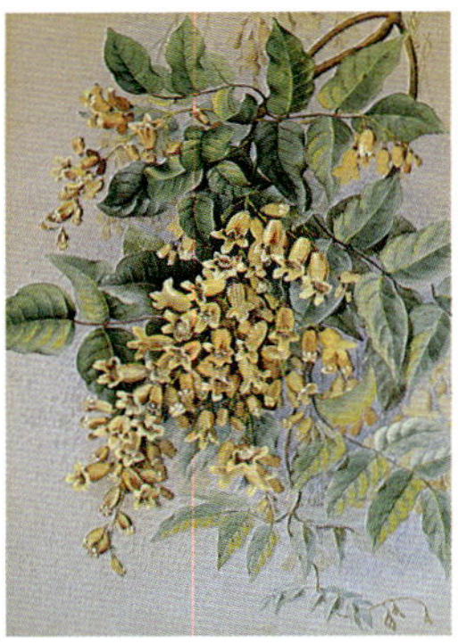

83

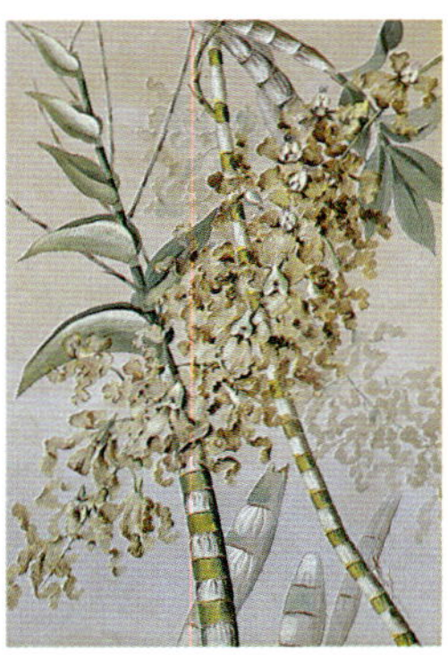

84

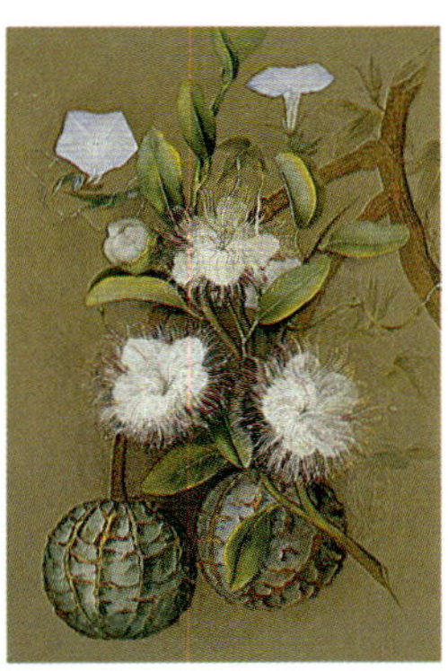

85

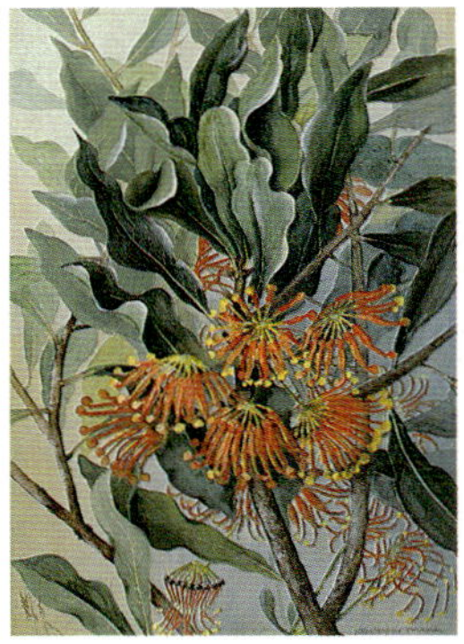

86

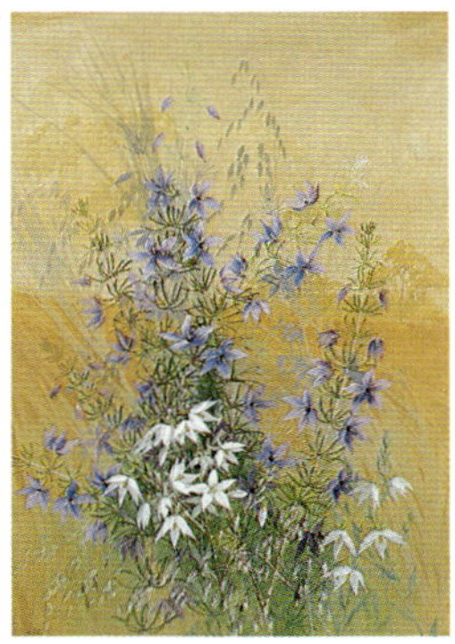

87

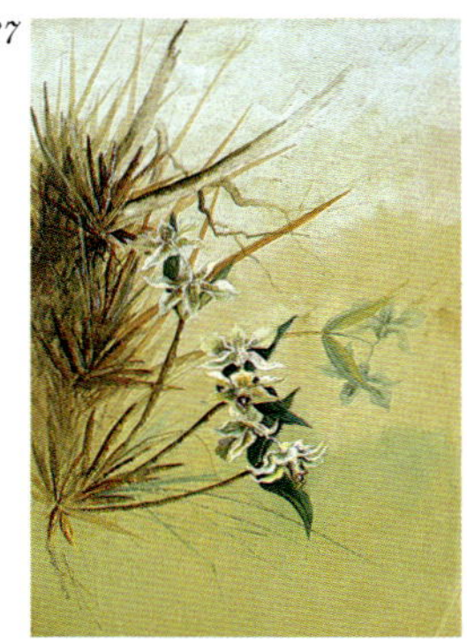

88

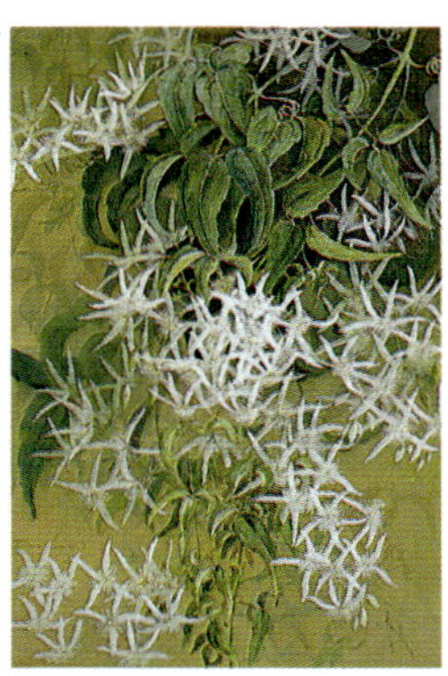

89

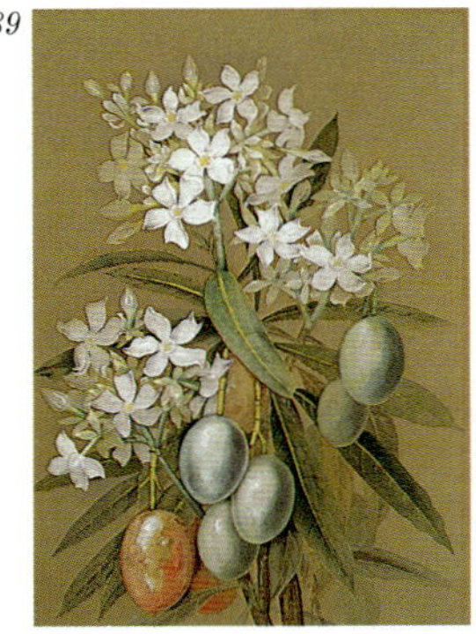

90

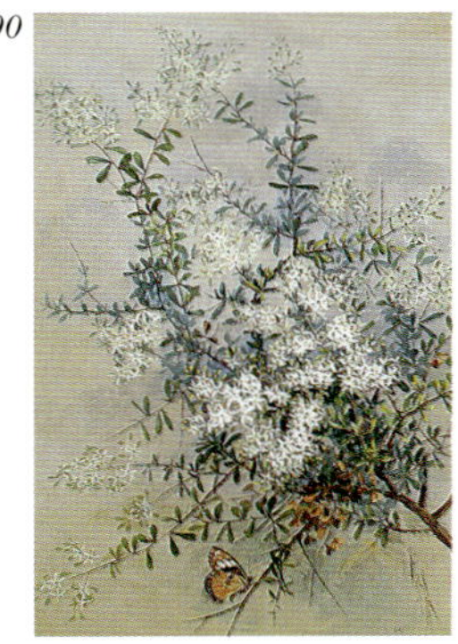

91

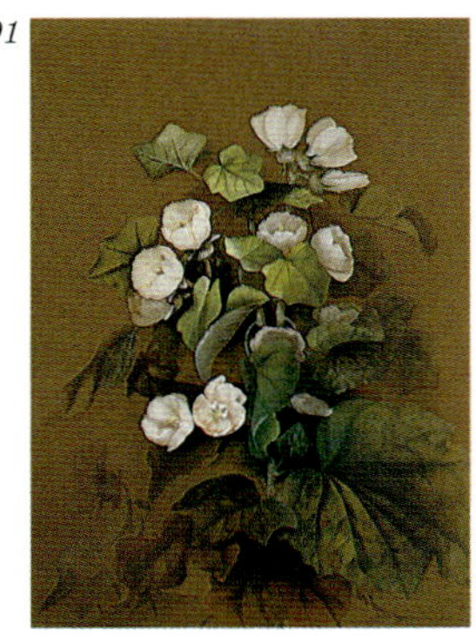

92

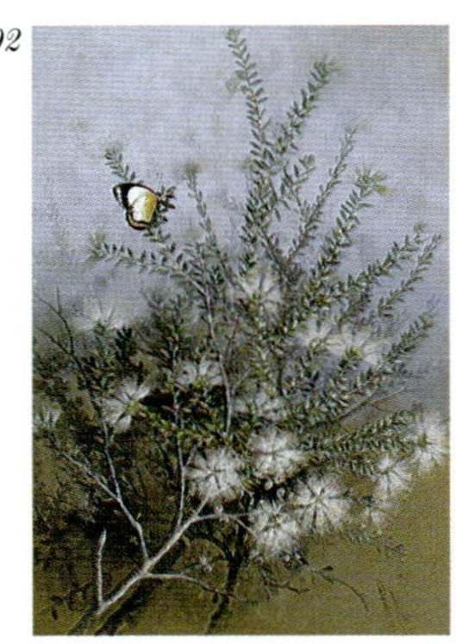

93

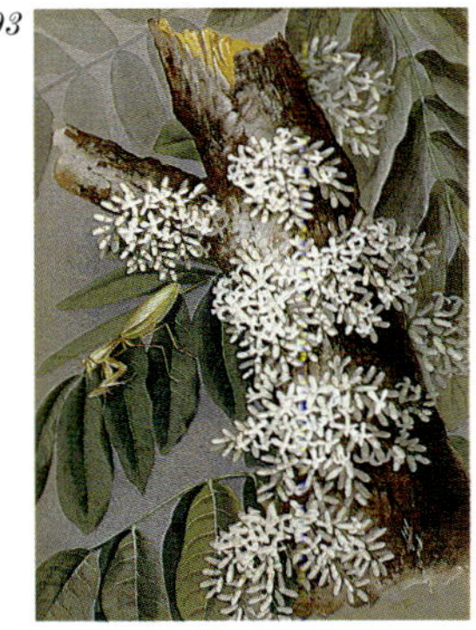

94

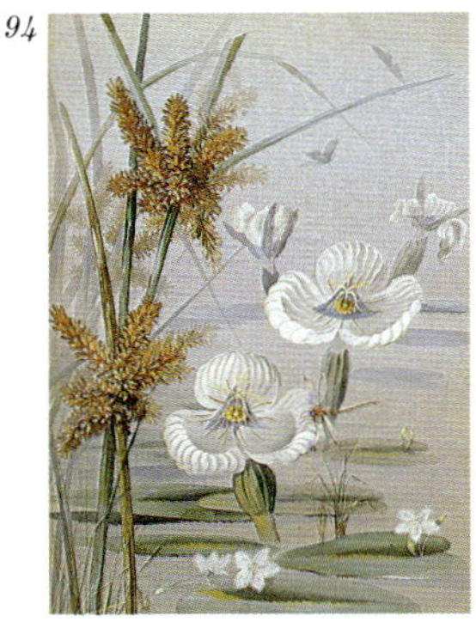

95

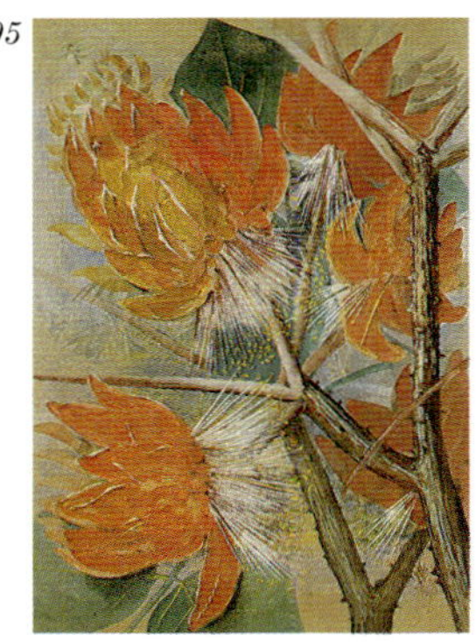

96

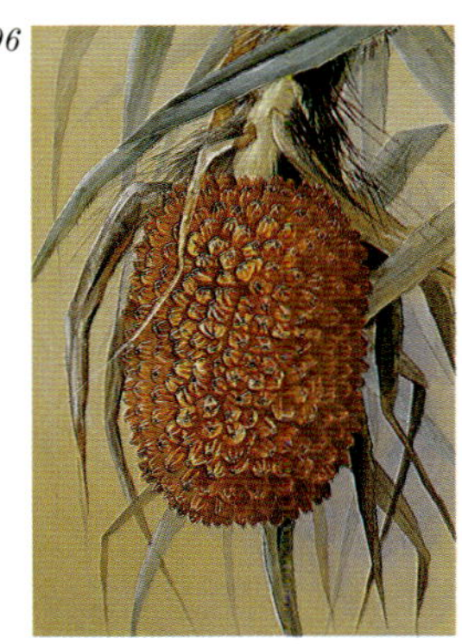

81
Banksia ericifolia L.f.
[Proteaceae]
HEATH-LEAVED BANKSIA
Introduced to Qld from NSW and widely cultivated

Painted: probably NSW (c.1910)

Titled by artist lower left corner "Banksia ericifolia"

Exhibited: Sydney 1910

82
Pandorea pandorana (Andrews) Steenis [Bignoniaceae]
WONGA VINE
Widespread in Qld except for SW; NSW, Vic.; New Guinea

Painted: Bloomfield River (1892)

Inscribed by artist reverse "22 / Bloomfield / 67"
Inscribed reverse "Tecoma sp. / Bignonia / NSW" (last two words crossed out)

Exhibited: Sydney 1910

83
Dendrobium discolor Lindley
[Orchidaceae]
GOLDEN ORCHID
E coast of Qld N of Gladstone

Painted: (1891-92)

Titled by artist reverse "Dendrobium undulatum / Var. Rowaniae"
Numbered reverse "86"

Exhibited: London 1896; Melbourne 1904; Sydney 1910, as *Dendrobium Rowaniae*

84
Foreground—*Capparis ornans* F.Muell. ex Benth.
[Capparaceae]
A NATIVE POMEGRANATE
E coast of Qld N of Bundaberg

Background vine—*Ipomoea nil* (L.) Roth [Convolvulaceae]
Coastal Qld and Torres Strait Is.; NT, WA

Painted: Jervis Is. (1891)

Inscribed by artist reverse "Jervis Island Torres Sts No. 2"
Titled by F.v. Mueller reverse "Capparis ornans FvM / Ipomoea hederacea Jacquin"

Exhibited: Sydney 1910

85
Stenocarpus sinuatus Endl.
[Proteaceae]
WHEEL OF FIRE
Rainforests of Qld; NSW

Painted: (1891-92)

Titled by artist lower right corner "Stenocarpus sinuatus / Queensland"
Numbered lower left corner "263"

Exhibited: Sydney 1910

86
Foreground—*Tetratheca thymifolia* Smith
[Tremandraceae]
E coast of Qld S of Bundaberg; NSW, Vic.

Background—*Avena* sp.
[Poaceae]

Painted: (1891-92)

Numbered by artist reverse "81"
Titled reverse "Tetretica" (sic.)
Numbered lower left corner "402"

Exhibited: Sydney 1910

87
Dendrobium atroviolaceum Rolfe
[Orchidaceae]
Introduced to Australia from New Guinea and widely cultivated

Painted: (1911)

Inscribed by C.T. White reverse "? New Guinea"

88
Clematis glycinoides DC.
[Ranunculaceae]
HEADACHE VINE
Rainforests of Qld; NSW, Vic.

Painted: (1891–92)

Inscribed reverse "Top / 407"

Exhibited: Sydney 1910

89
Cerbera manghas L.
[Apocynaceae]
MILKWOOD; NATIVE OLEANDER
Strand forests of Qld N of Gympie; New Guinea and other SW Pacific Is.

Painted: Somerset (1891–92)

Titled by artist on original hinge "Native oleander"
Inscribed by artist reverse "Somerset / No. 3"
Numbered lower left corner "227" and reverse "48"

Exhibited: London 1896; Melbourne 1904; Sydney 1910

Possibly referred to: *Flower-Hunter*, p.138

90
Bursaria spinosa Cav.
[Pittosporaceae]
PRICKLY PINE; BLACKTHORN
E coast of Qld; NSW, SA, Vic., Tas.

Insect—*Hypolimnas misippus* (L.) [Nymphalidae]
DANAID EGGFLY BUTTERFLY

Painted: (1891-92)

Titled reverse "Native Privet"
Numbered lower right corner "416"

Exhibited: Sydney 1910

91
Malva sp. [Malvaceae]
Possibly introduced to Australia as an ornamental

Painted: probably Mackay 1887

Numbered reverse "Q9"

92
Melaleuca sieberi Schauer
[Myrtaceae]
Coastal SE Qld; NSW

Insect—*Delias mysis* (Fabricius) [Pieridae]
UNION JACK BUTTERFLY

Painted: (1891–92)

Numbered by artist reverse "75"
Titled reverse "Meleluca" (sic.)

Exhibited: Sydney 1910

93
Dysoxylum parasiticum (Osbeck) Kosterm. [Meliaceae]
BUSH WALNUT
Rainforests of Qld N of Ingham

Insect—*Hierodula* sp.
[Mantidae]
A PREYING MANTIS

Painted: Somerset (1891-92)
Inscribed by artist reverse "36 / Somerset"

Exhibited: Possibly Sydney 1910

94
Large flowers—*Ottelia ovalifolia* (R.Br.) Rich. [Hydrocharitaceae]
WATER POPPY
Mainly E Qld; NSW, NT, SA, Vic.

Small flowers below—*Nymphoides indica* (L.) Kuntze
[Menyanthaceae]
WATER SNOWFLAKE
E Qld; NSW, NT, WA; Asia, Africa, America

Sedge on left—*Cyperus alopecuroides* Rottb.
[Cyperaceae]
A SEDGE
C Qld; NT

Insect—Indeterminate species
[Zygoptera]
A DAMSEL FLY

Painted: after the original, Mackay 1887

Inscribed by artist reverse "Cyperus / Limnathemum Indicum / Ottelia ovalifolia 2 / 77"

Reproduced: *Press Opinions Mrs F.C. Rowan's Watercolour Flower Drawings*, London, (1905)

Exhibited: London 1896; Sydney 1910

95
Erythrina crista-galli L.
[Fabaceae]
A CORAL TREE
Introduced to Australia and widely cultivated; Pantropical

Painted: Murray Is. (1892)

Numbered by artist reverse "2"
Titled by F.v. Mueller reverse "Erythrina"
Inscribed reverse original backing board "Murry Island / Erythrina" (sic.)
Numbered lower right corner "213"

Exhibited: London 1896; Sydney 1910

Referred to: *Flower-Hunter*, pp.163–64

96
Pandanus tectorius var. *pedunculatus* (R.Br.) Domin
[Pandanaceae]
SCREWPINE
E coast of Qld S of Rockhampton; NSW; SW Pacific Is.

Painted: after the original, Mackay 1887

Inscribed by artist reverse "Pandanus pedunculatus / Queensland / 80"

Exhibited: Sydney 1910

97
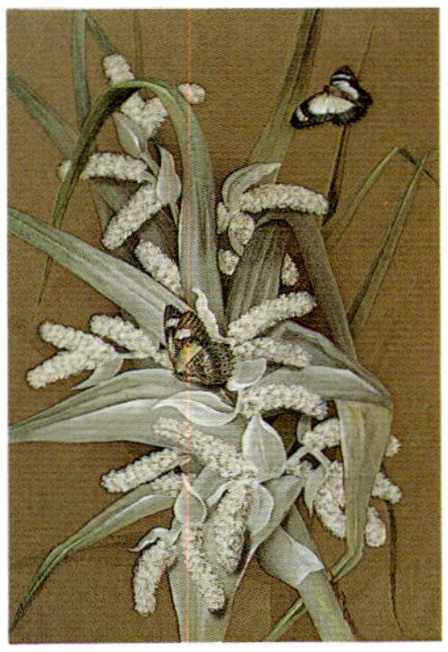

98
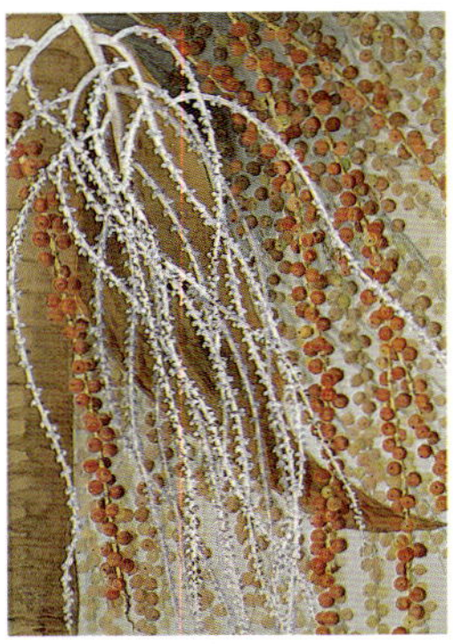

99
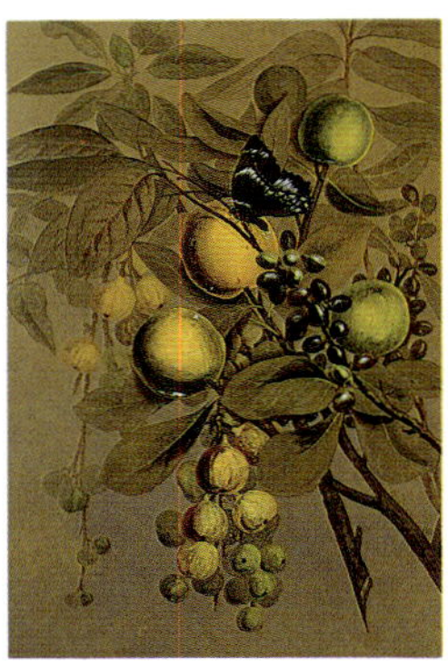

100
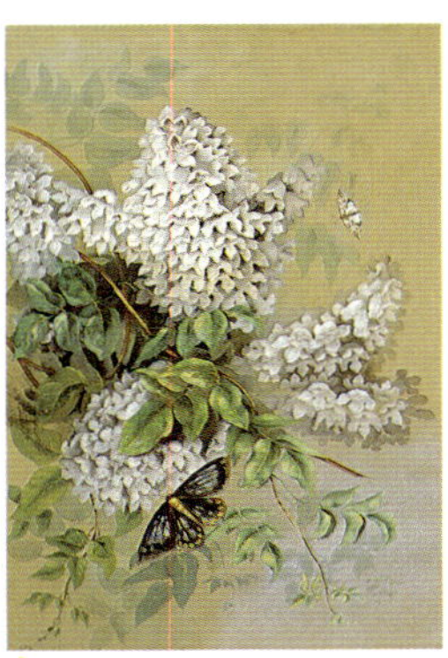

101
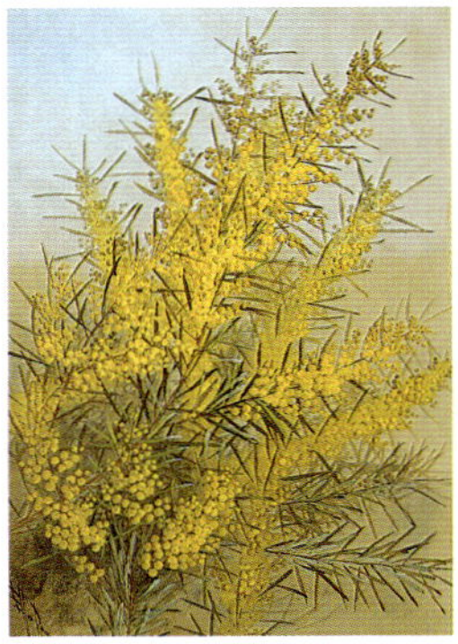

102
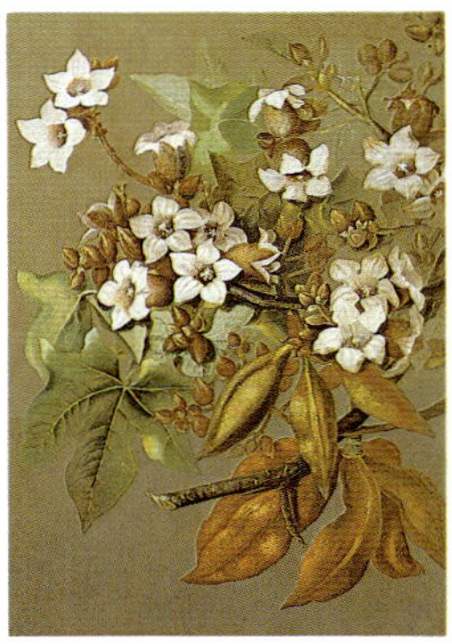

103
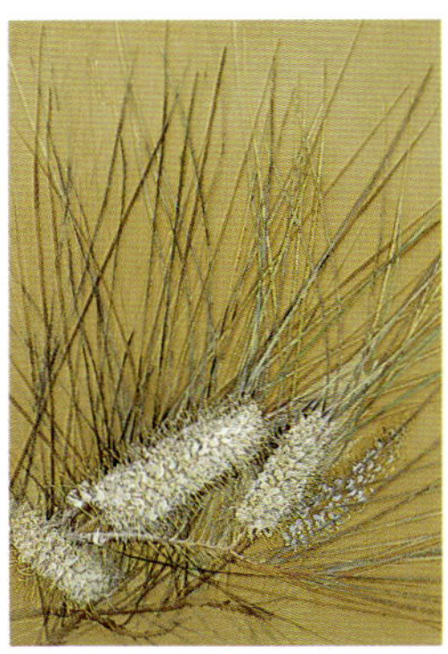

104
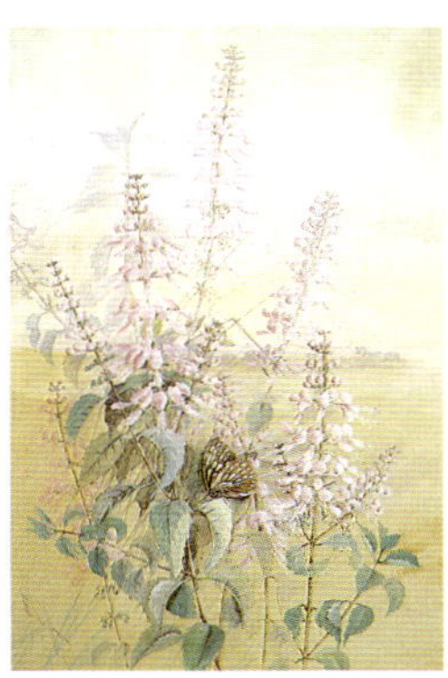

105
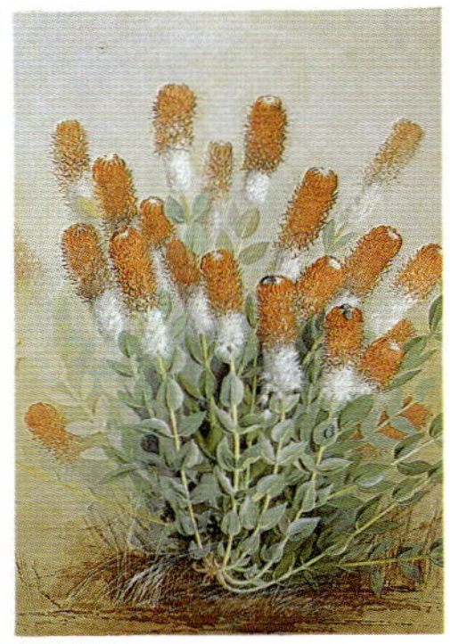

106
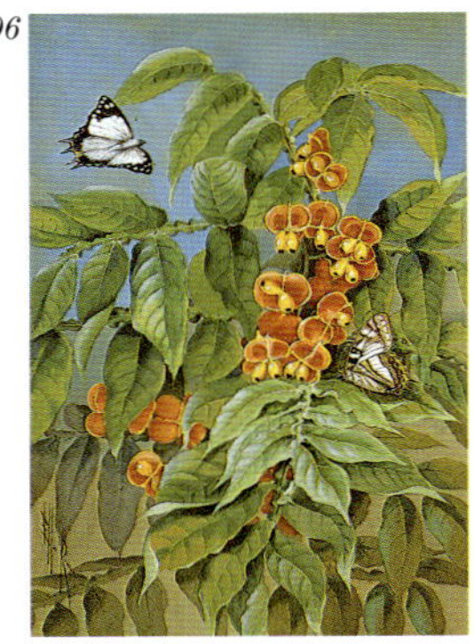

107
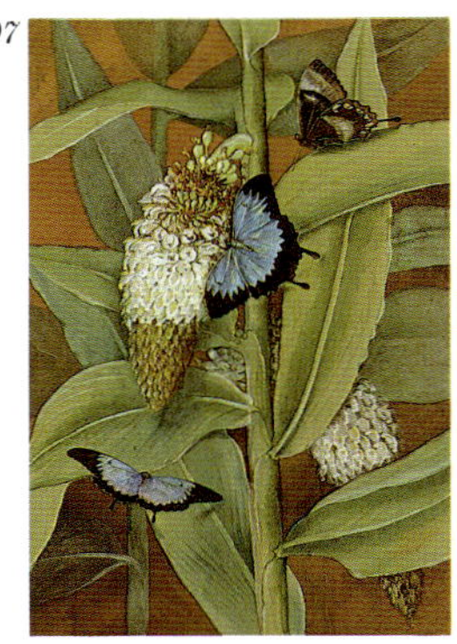

108
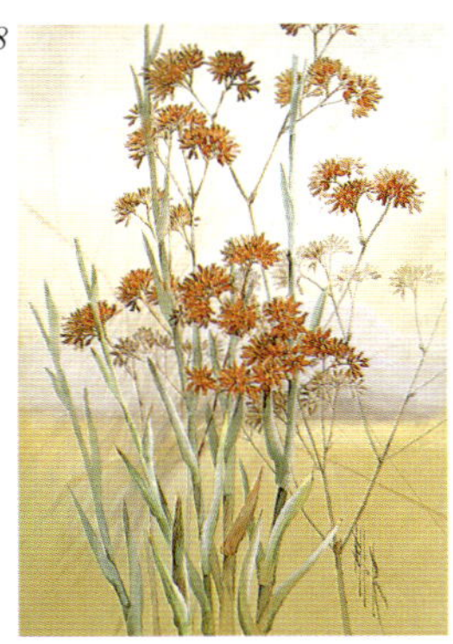

109
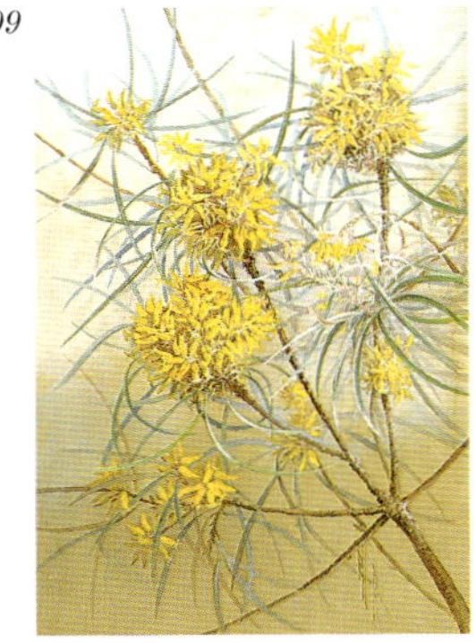

110
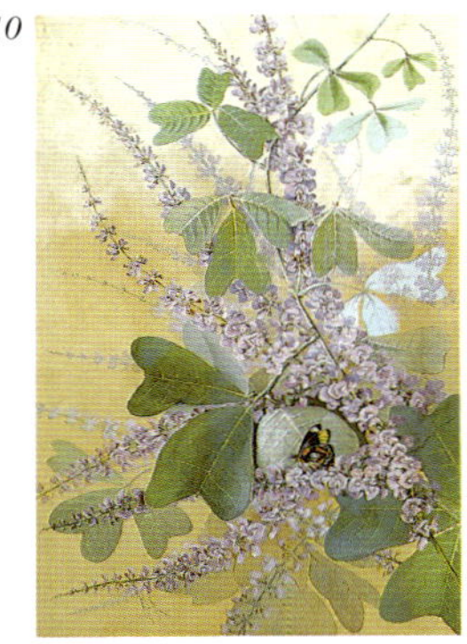

111
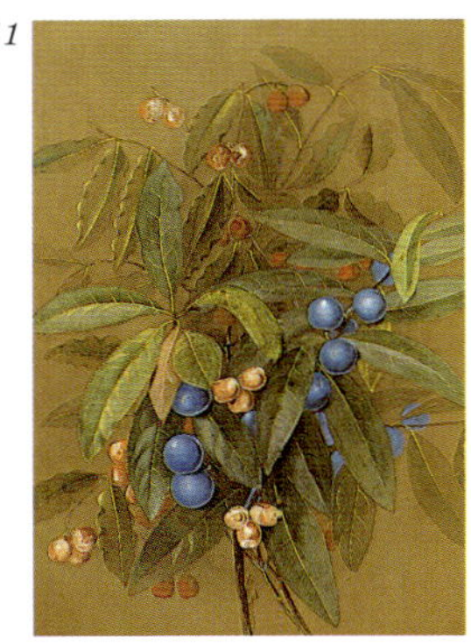

112
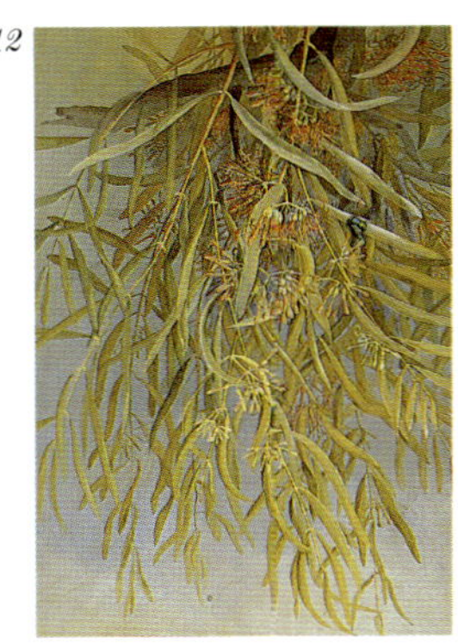

97
Pandanus monticola F.Muell.
[Pandanaceae]
SCREWPINE
E coast of Qld N from Townsville area

Insects—*Cethosia cydippe* L. [Nymphalidae]
RED LACEWING BUTTERFLY

Painted: Johnstone River 1887

Titled by F.v. Mueller reverse "Pandanus monticola, F.v. Mueller"
Inscribed reverse "Q53 Johnstone River"
Numbered lower right corner "419"

Exhibited: London 1896; Sydney 1910

98
Ptychosperma elegans R.Br.
[Arecaceae]
ELEGANT PALM
E coast of Qld N from Townsville area; NT

Painted: (?1891-92)

Titled by the artist lower left corner and reverse "Ptychosperma Alexandrae Palm / Queensland"

99
Large fruits—*Planchonella pohlmaniana* (F.Muell.) Dubard [Sapotaceae]
YELLOW BOXWOOD; ENGRAVER'S WOOD
E coast of Qld; NSW

Pendulous fruits below—*Ficus racemosa* L. [Moraceae]
CLUSTER FIG
E coast of Qld N from Maryborough

Small dark shiny fruits—*Cryptocarya* sp. [Lauraceae]

Insect—*Papilio aegeus* Donovan [Papilionidae]
ORCHARD SWALLOWTAIL BUTTERFLY

Painted: Mackay 1887

Titled by F.v. Mueller reverse "1. Ficus / 2. Sideroxylon / 3.—"
Inscribed reverse "Q26 Mackay"
Numbered lower left corner "137"

Exhibited: Sydney 1910

100
Pandorea pandorana (Andrews) Steenis [Bignoniaceae]
WONGA VINE
Widespread in Qld except for SW; NSW, Vic.; New Guinea

Insects—*Dysphania fenestrata* (Swainson) [Geometridae]
A DAY-FLYING MOTH
Utetheisa sp. [Arctiidae]
A SALT AND PEPPER MOTH

Painted: (1891-92)

Inscribed by artist reverse "78 / Queensland"
Titled by F.v. Mueller reverse "Tecoma australis, R. Brown"
Numbered lower left corner "434"

Exhibited: London 1896; Melbourne 1904

101
Acacia fimbriata G.Don
[Mimosaceae]
BRISBANE GOLDEN WATTLE
E coast of Qld S from Townsville area; NSW

Painted: (1911)

Reproduced: Qld Department of Public Instruction. *The School Paper*, for Classes 5 and 6, New Series, Vol. 4, July 1913

102
Brachychiton discolor F.Muell.
[Sterculiaceae]
LACEBARK TREE; BUSH KURRAJONG
E coast of Qld S of Mackay; NSW

Painted: (1891–92)

Titled by F.v. Mueller reverse (illeg.) "Hook."

103
Grevillea striata R.Br.
[Proteaceae]
BEEFWOOD
Widespread in Qld except for far SW; NSW, NT, WA, SA

Painted: (1911)

Numbered by artist reverse "41"
Titled by C.T. White reverse "Grevillea sp."

104
Orthosiphon aristatus (Blume) Miq. [Lamiaceae]
WHISKER PLANT
E coast of Qld N of Ingham; New Guinea and other SW Pacific Is., Malaya, India

Insect—*Tirumala hamata* (Macleay) [Nymphalidae]
BLUE TIGER BUTTERFLY

Painted: (1911)

Numbered by artist reverse "36"
Titled by C.T. White reverse "Orthosiphon / stamineus"

105
Pimelea decora Domin
[Thymelaeaceae]
FLINDERS POPPY
C and NW Qld

Insects—possibly *Chalcopteroides* [Tenebrionidae]

Painted: (1911)

Numbered by artist reverse "52"
Titled by C.T. White reverse "Pimelea / haematostachya"

Similar work reproduced: Bailey, plate 14; Tardent, p. 37; *Muses Magazine*, July and September 1928

106
Harpullia alata F.Muell.
[Sapindaceae]
Rainforests of Qld S of Maryborough; NSW

Insects—*Polyura pyrrhus* (L.) [Nymphalidae]
TAILED EMPEROR BUTTERFLY

Painted: (1911)

Numbered by artist reverse "47"
Titled by C.T. White reverse "Harpullia / alata"

107
Alpinia racemigera F.Muell.
[Zingiberaceae]
Cape York Peninsula in Qld

Insects—*Papilio ulysses* L. [Papilionidae]
ULYSSES BUTTERFLY

Painted: (1911)

Numbered by artist reverse "51"

Similar work reproduced: Bailey, plate 16; Tardent, p. 41; *Muses Magazine*, December 1927

108
Haemodorum coccineum R.Br.
[Haemodoraceae]
SCARLET-FLOWERED BLOOD-ROOT
Coastal Qld N of Proserpine

Painted: (1911)

Numbered by artist reverse "32"
Titled by C.T. White reverse "Blood Root / Haemodorum / coccineum"

109
Persoonia falcata R.Br.
[Proteaceae]
GEEBUNG
E coast of Qld; NT

Painted: (1911)

Numbered by artist reverse "7"
Titled by C.T. White reverse "Persoonia / falcata"

110
Vandasina retusa (Benth.) Rauschert [Fabaceae]
E coast of Qld N from Innisfail; New Guinea

Insect—*Delias nigrina* (Fabricius) [Pieridae]
COMMON JEZABEL BUTTERFLY

Painted: (1911)

Numbered by artist reverse "8"
Titled by C.T. White reverse "Hardenbergia / retusa"

Similar work reproduced: Bailey, plate 5; Tardent, p. 33; *Muses Magazine*, January 1928

111
Large fruits—*Elaeocarpus grandis* F.Muell.
[Elaeocarpaceae]
BLUE QUANDONG; SILVER QUANDONG
Rainforests of Qld; NSW

Small fruits—*Acmena hemilampra* (F.M. Bailey) Merr. & Perry subsp. *hemilampra* [Myrtaceae]
BLUE SATIN ASH
Rainforests of Qld; NSW

Painted: Johnstone River 1887

Titled by F.v. Mueller reverse "1. Fruits? / 2. Elaeocarpus grandis F.v. Mueller"
Inscribed reverse "Q64 Pioneer River" originally "Johnstone River"

Exhibited: Probably Melbourne 1888

112
Amyema miquelii (Miq.) Tieghem [Loranthaceae]
A MISTLETOE
Widespread in Qld except for SW; NSW, NT, WA, SA, Vic.

Insect—*Rutilia* sp. [Tachinidae]
A PARASITIC FLY

Painted: (1911)

Numbered by artist reverse "37"
Titled by C.T. White reverse "Loranthus / pendulus"

113

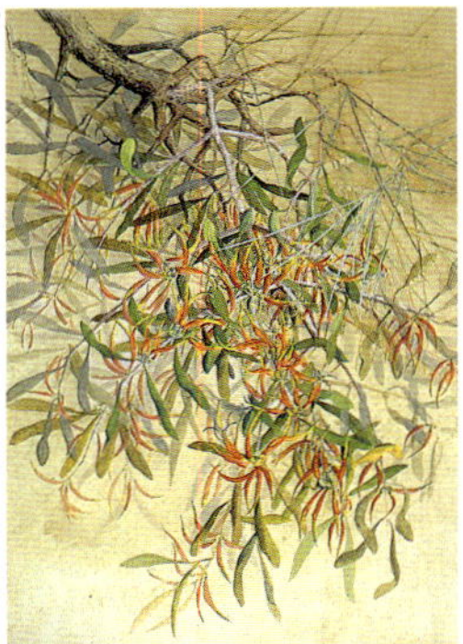

114

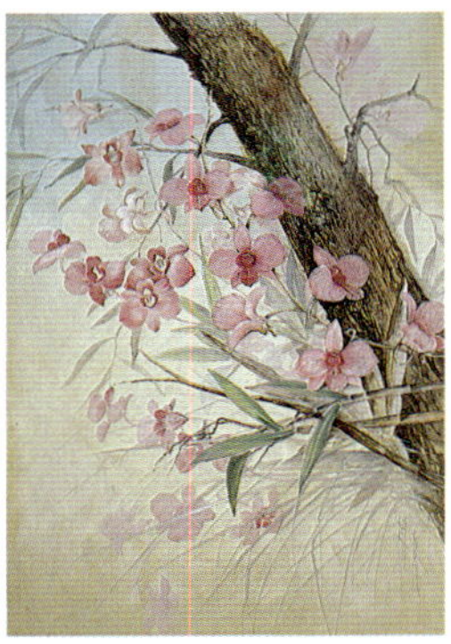

115

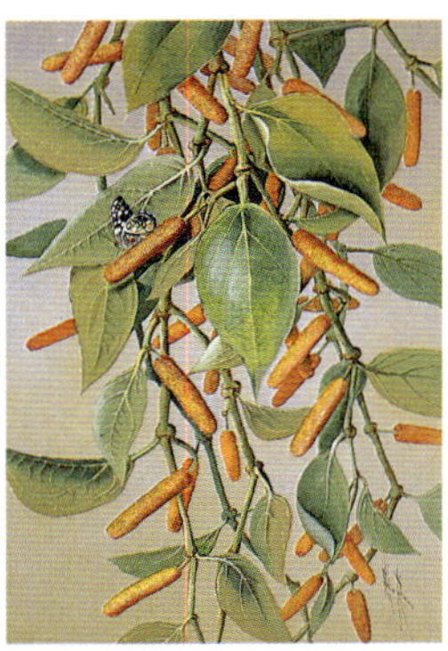

116

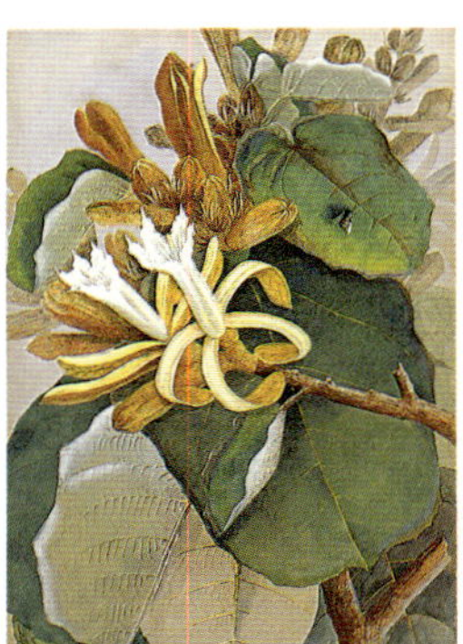

117

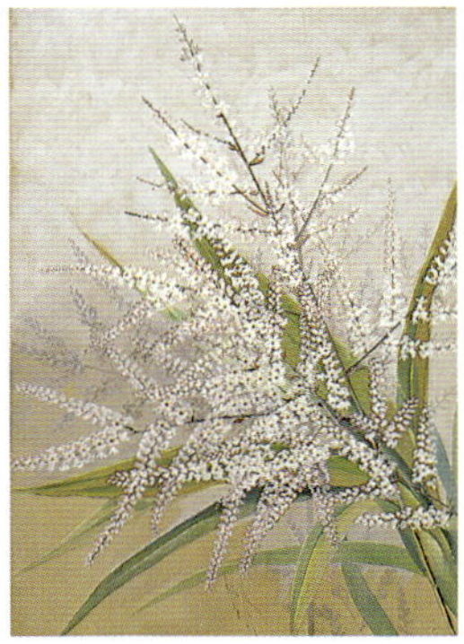

118

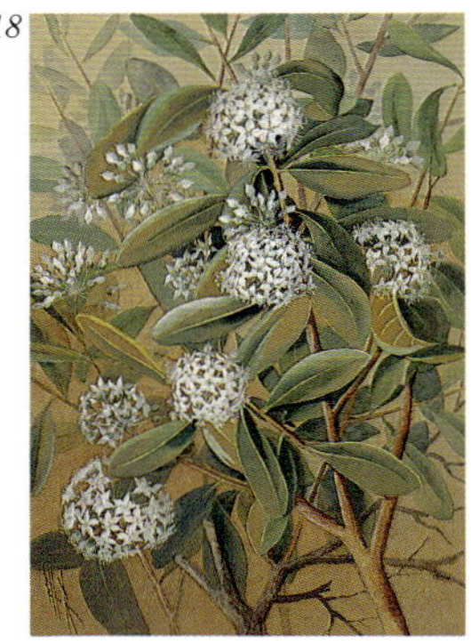

119

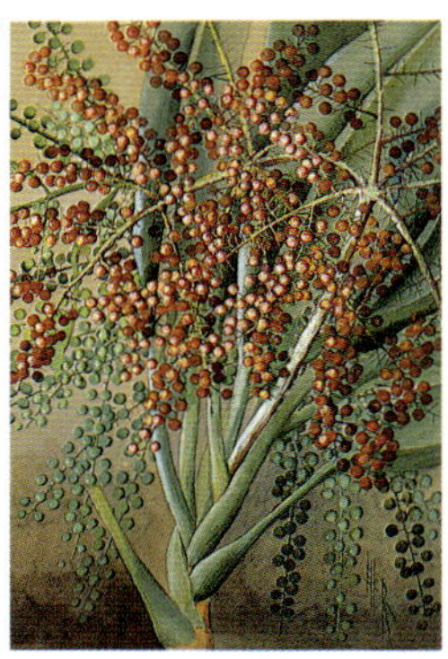

120

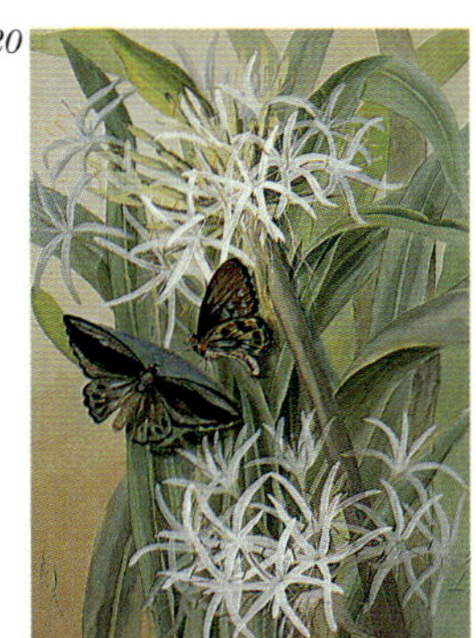

121

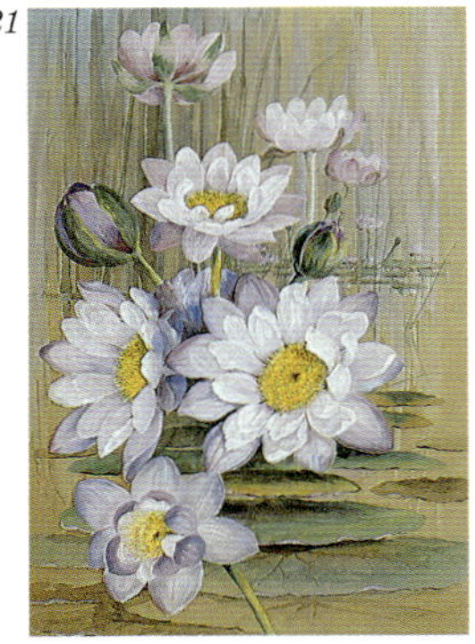

122

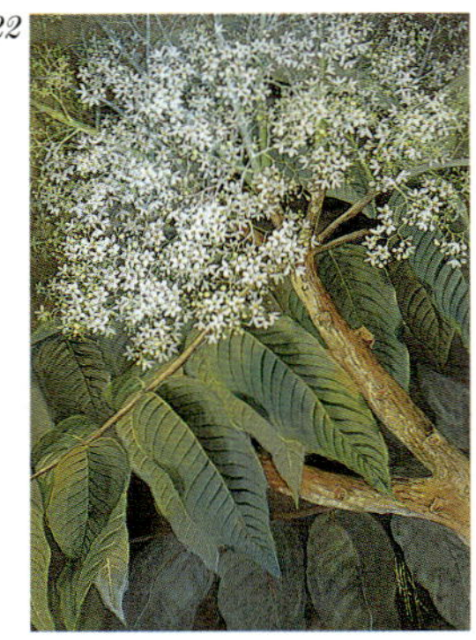

123

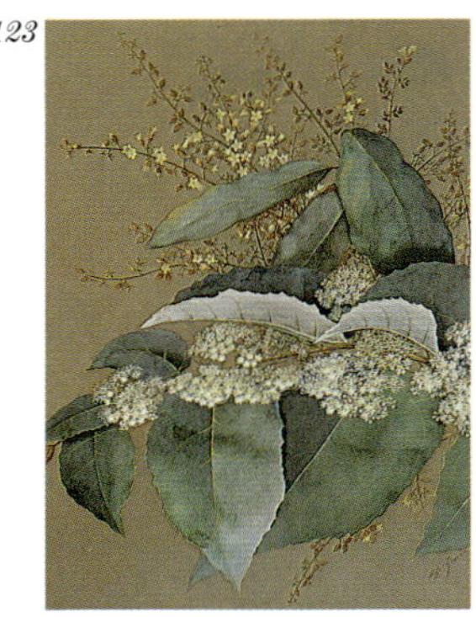

124

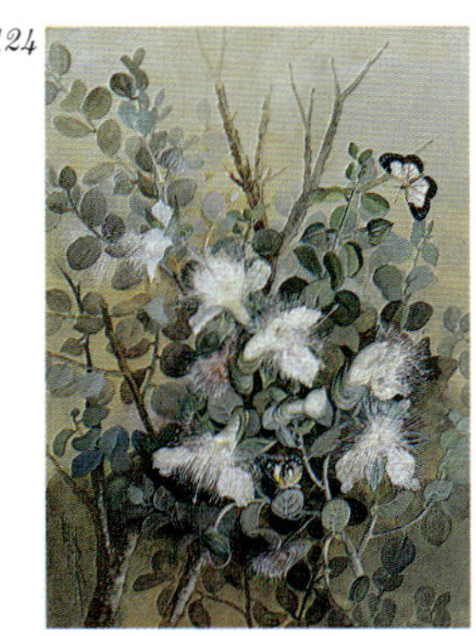

125

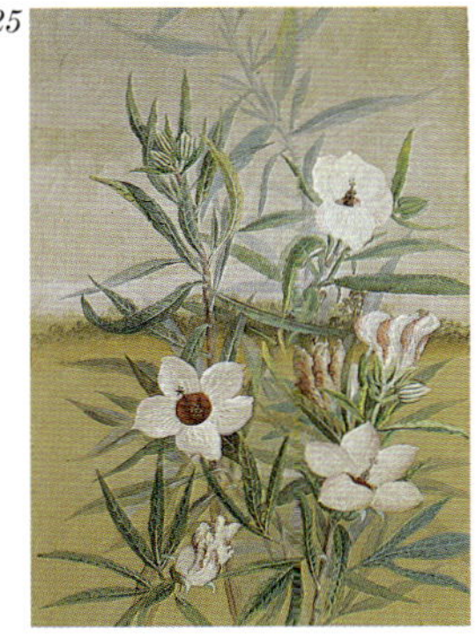

113
Amyema pendulum (Sprengel) Tieghem [Loranthaceae]
A MISTLETOE
C coast and SE Qld; NSW, SA, Vic.

Painted: (1911)

Numbered by artist reverse "42"
Titled by C.T. White reverse "Loranthus sp."

114
Dendrobium bigibbum Lindley [Orchidaceae]
COOKTOWN ORCHID
Cape York Peninsula in Qld and Torres Strait Is., widely cultivated

Painted: (1911)

Titled by C.T. White reverse "Dendrobium / bigibbum"

115
Piper mestonii F.M. Bailey [Piperaceae]
A NATIVE PEPPER
Rainforests of Qld

Insect—*Papilio demoleus* L. [Papilionidae]
CHECKERED SWALLOWTAIL BUTTERFLY

Painted: (1911)

Numbered by artist reverse "21"
Titled by C.T. White reverse "Piper Mestoni"

Similar work reproduced: Bailey, plate 12; Tardent, p. 53; *Muses Magazine*, January 1928

116
Pterospermum acerifolium Willd. [Sterculiaceae]
Introduced to Australia from SE Asia and cultivated

Insect—*Rutilia* sp. [Tachinidae]
A PARASITIC FLY

Painted: (1911)

Numbered by artist reverse "10"
Titled by C.T. White reverse "Pterospermum / acerifolium"

117
Cordyline fruticosa (L.) A.Chev. [Agavaceae]
Cape York Peninsula in Qld; cultivated in NSW; SE Asia, Pacific Is.

Painted: (1911)

Numbered by artist reverse "11"
Titled by C.T. White reverse "Cordyline / terminalis / a Lily Palm"

118
Aegiceras corniculatum (L.) Blanco [Myrsinaceae]
RIVER MANGROVE
Mangrove forests of coastal Qld; NSW, NT, WA

Painted: (1911)

Numbered by artist reverse "65"
Titled by C.T. White reverse "Aegiceras / majus / River Mangrove"

119
Cordyline cannifolia R.Br. [Agavaceae]
Rainforests of Qld N of Townsville; NT

Painted: (1911)

Numbered by artist reverse "12"
Titled by C.T. White reverse "Cordyline / terminalis / Lily-palm / fruit"

120
Crinum pedunculatum R.Br. [Liliaceae]
RIVER LILY
Coastal Qld; NSW

Insects—*Ornithoptera priamus* (L.) [Papilionidae]
BIRDWING BUTTERFLY

Painted: (1911)

Numbered by artist reverse "61"
Titled by C.T. White reverse "Crinum / pedunculatum"

121
Nymphaea gigantea J.D. Hook. [Nymphaeaceae]
GIANT BLUE WATERLILY
Swamps and waterways of coastal Qld; NSW, NT

Painted: (1911)

Numbered by artist reverse "58"
Titled by C.T. White reverse "Nymphaea / gigantea"

122
Flindersia schottiana F.Muell. [Rutaceae]
BUMPY ASH; SILVER ASH
Rainforests of Qld; NSW; New Guinea

Painted: (1911)

Numbered by artist reverse "28"
Titled by C.T. White reverse "Flindersia / pubescens"

Similar work reproduced: Bailey, plate 4

123
Foreground - *Commersonia bartramia* (L.) Merr. [Sterculiaceae]
BROWN KURRAJONG
E coast of Qld; New Guinea and other SW Pacific Is., SE Asia, Philippines, Indonesia

Background—?

Painted: (1891-92)

Exhibited: Sydney 1910

124
Capparis spinosa var. *nummularia* (DC.) F.M. Bailey [Capparaceae]
Inland and coastal tropical Qld; NT, WA

Insects—*Anaphaeis java* Sparrman [Pieridae]
CAPER WHITE BUTTERFLY

Painted: (1911)

Numbered by artist reverse "3"
Titled by C.T. White reverse "Capparis / nummularia"

125
Hibiscus meraukensis Hochr. [Malvaceae]
MERAUKE HIBISCUS
Widespread in Qld except for far SW; NT, WA; New Guinea

Painted: (1911)

Indexes — *Plants and animals depicted in the paintings are indicated by cat. nos.*

Plant Scientific Names

Plant Common Names

Systematic List of Animals

APPENDIX 1: EXHIBITIONS

Queensland paintings by Ellis Rowan were exhibited as follows during her lifetime:

Date	*Exhibition*	*Select Reviews*
1888	Centennial International Exhibition, Melbourne, in the Queensland Court 1888. Awarded first order of merit with special mention and a gold medal	*(Argus Exhibition Supplement (Melbourne)*, 9 October *Boomerang* (Brisbane), 15 December 1888, p. 13
1891 30 October — "for about a fortnight"	Ellis Rowan exhibition, Buxton's Art Gallery, Swanston St, Melbourne	*Age* (Melbourne), 29 October 1891, p. 5; *Argus*, 29 October 1891, p. 5, and 30 October 1891, p. 5
1893	World's Columbian Exposition, Chicago, in the hall of "Australia House", the New South Wales Government Building. Awarded a gold medal	*Australasian Builder and Contractors' News*, 29 July 1893, p. 56
1896 20 April — "for three or four weeks"	Ellis Rowan exhibition, Dowdeswell Galleries, New Bond St, Mayfair, London	*Press Opinions of Mrs F.C. Rowan's Watercolour Flower Drawings*, Stenlake and Simpson, London, (1905); *Queenslander*, 20 June 1896, p. 1171
1902 21 January — 1 February	Ellis Rowan exhibition, Clausen's Gallery, Fifth Ave, New York	*Argus*, 9 December 1904, p. 11, and 10 December 1904, p. 17
1904 1 — 7 December	Ellis Rowan exhibition, Scourfield Chambers, Collins St, Melbourne	*Age*, 6 July 1909, p. 7
1909 5 — 31 July	Ellis Rowan exhibition, Bernard's Gallery, Collins St, Melbourne	*Sydney Mail*, 10 August 1910, p. 54; *Sydney Morning Herald*, 10 August 1910, p. 5, and 23 August 1910, p. 10
1910 25 July — 13 August	Ellis Rowan exhibition, Angus and Robertson's Gallery, Castlereagh St, Sydney	*Brisbane Courier*, 6 August 1912, p. 9, and 14 September 1912

1912 5 August — c. 14 September	Ellis Rowan exhibition, old Town Hall, Queen St, Brisbane	Courier Home Circle, p.4; *Daily Mail* (Brisbane), 6 August 1912, p.2; *Queensland Figaro*, 10 September 1912, p. 4; *Ellis Rowan Cuttings Book*, pp. 38-42, MS2203, NLA.
1914 4 December — "for a fortnight"	Ellis Rowan exhibition, Fine Art Society's Gallery, Collins St, Melbourne	*Argus*, 5 December 1914, p. 9
1915 25 March — "for three weeks"	Ellis Rowan exhibition, Anthony Hordern's Gallery, New Palace Emporium, Brickfield Hill, Sydney	*Sydney Morning Herald*, and 7 April 1915, p. 5 26 March 1915, p. 8,
1918 6 — 19 November	Ellis Rowan exhibition, Fine Art Society's Galleries, Alfred Place, Melbourne	*Argus*, 5 November 1918, p. 6, and 7 November 1918, p. 8.
1919	Ellis Rowan touring exhibition, shown widely in the United States of America, including at the Art Gallery of Stanford University, California, and at the American Museum of Natural History, New York	
1920 2 — 31 March	Ellis Rowan exhibition, Anthony Hordern's Gallery, New Palace Emporium, Brickfield Hill, Sydney	*Sydney Morning Herald*, 2 March 1920, p. 8, and 3 March 1920, p. 10.

E N D N O T E S

Biography

1. Joan Kerr, Entry on Ellis Rowan in *200 Australian Women: A Redress Anthrology*, Women's Redress Press, Sydney, 1988, p. 53.

2. Letter, Robert Bruce to Mrs (Professor) Alfred Cort Haddon, 4 September 1892, packet 1006, *Haddon Collection*, Cambridge University Library. I am indebted to Dr Ruth Kerr for finding this letter.

3. 'The Victorian Artists' Society, The awards of the fine arts jury of the Exhibition', *Argus* (Melbourne), 23 January 1889, p. 5.

4. Ellis Rowan, 'Making of an Art Collection', *Oarswoman* (Brisbane), December 1912, p. 10.

5. M. Hirst Browne, 'Some Australian Women', *Illustrated Sydney News*, 9 May 1891, p. 10.

6. Transcript of a prospectus issued by the Picturesque Atlas Publishing Company, n.d., MS422, NLA.

7. See *Australian Dictionary of Biography*, vol. 5, 1851-1890, pp. 306-08.

8. Ellis Rowan, 'An Australian Artist's Adventures', *New Idea*, 6 February 1905, p. 714.

9. Margaret Hazzard, *Australia's Brilliant Daughter Ellis Rowan: Artist, Naturalist, Explorer 1848-1922*, Greenhouse Publications, Melbourne, 1984, p. 109.

10. Letter, Frederic Rowan to the Queensland Premier Sir Samuel Griffith, 19 August 1887, 1887/6819, COL/A514, QSA.

11. This excuse was given in the preface of *Flower-Hunter* and repeated in contemporary articles about Ellis Rowan.

12. Hazzard, p. 108.

13. For example, Ellis disembarked at "unbearably hot" Murray Island wearing her gloves *(Flower-Hunter*, p. 156).

14. Outline of Ellis Rowan's career accompanying her exhibition at New York in 1902, MS2206, NLA.

15. *Flower-Hunter*, pp. 119-22.

16. Maie Casey, *An Australian Story 1837-1907*, Michael Joseph, London, 1962, p. 106.

17. Letter, Frederic Rowan to the Queensland Premier, op. cit.

18. Ellis was not the first woman to travel extensively in Queensland in search of botanical specimens. The German naturalist Amalie Dietrich spent the years 1863-72 collecting botanical and other specimens for a museum in Hamburg.

19. *Flower-Hunter*, p. 119.

20. An orchid which she discovered was named '*Dendrobium undulatum* var. *Rowanae*' (cat. no. 83) after her by von Mueller, but later reverted to an earlier name.

21. 'Our Portraits. An Australian Flower Painter', *Ellis Rowan Cuttings Book*, MS2203, p. 13, NLA. Ellis had a painful encounter with a nettle tree during her 1887 visit (*Ellis Rowan Letters from Queensland*, 1891, third letter, p. 22, held by the artist's family).

22. 'Death of Captain F. C. Rowan', *Argus*, 12 December 1892, p. 6.

23. Joseph Dana Miller, 'Mrs Ellis Rowan, the World's Greatest Painter of Wildflowers', *Ellis Rowan Cuttings Book*, p. 7.

24. Letter, from Windsor Castle to Ellis Rowan, 17 May 1895, MS2206, NLA.

25. 'Paintings of Australian Flora', *Queenslander*, 20 June 1896, p. 1171.

26. 'Notes and Notices', *Australasian* (Melbourne), 4 January 1896, p. 32.

27. Letter, Ellis Rowan to the Queensland Agent-General, London, 19 August 1895, 1895/11895, COL/128, QSA. The Imperial Institute was opened in 1893 as a permanent museum of the resources of the British Empire.

28. Casey, pp. 107, 132.

29. William Moore, *The Story of Australian Art*, Angus and Robertson, Sydney, 1934, vol. 2, p. 33.

30. For example, 'The Rowan Collection', Letter to the Editor by Julian Ashton, *Sydney Morning Herald*, 3 December 1921, p. 13; 'A Woman's Letter', *Bulletin*, 1 February 1923, p. 26.

31. 'Mrs Rowan's Pictures. To Buy, or Not to Buy? "Disastrous Art". Opinions of Leading Artists', *Sun* (Sydney), 17 January 1923, p. 6.

Flower-Hunting in Queensland

1887

1. Shipping departures, *Queenslander*, 3 September 1887, p. 382.

2. Letter, Frederic Rowan to the Queensland Premier, op. cit.

3. In *Flower-Hunter* (pp. 12-16, 21-22), Ellis called the property 'The Rocks', but her descriptions match the well-known property of Charles and Edward Rawson at Mirani.

4. *Flower-Hunter*, pp. 13-14.

5. This painting attracted attention when displayed in 1888 in the Centennial International Exhibition, Melbourne *(Boomerang* (Brisbane), 15 December 1888, p. 13).

6. *Flower-Hunter*, pp. 17-19.

7. ibid, p. 23.

8. Shipping departures, *Townsville Herald*, 22 October 1887, p. 22.

9. The Herbert River Pastoral and Agricultural Association's annual show at Ingham was held on 12 November in 1887.

10. *Flower-Hunter*, pp. 24-28.

11. ibid, pp. 28-30.

12. ibid, p. 32.

13. ibid, pp. 32-38.

14. Shipping arrivals, *Townsville Daily Bulletin*, 27 December 1887, p. 2.

15. Shipping arrivals, *Brisbane Courier*, 2 January 1888, p. 4.

16. 'Queensland', *Argus Exhibition Supplement*, 9 October 1888, p. 1 — this probably refers to cat. nos 66, 111.

17. 'The Painter of the Picture. Mrs Ellis Rowan', *Australian Town and Country Journal*, 22 December 1888, p. 1238.

18. *Catalogue of Exhibits in the Queensland Court, Centennial International Exhibition, Melbourne, 1888*, Warwick and Sapsford, Brisbane, 1888, p. 65, item 556. This erroneously describes Ellis Rowan's paintings as in oils. The paintings are not listed by title.

19. 'The Art Section — Queensland', *Argus Exhibition Supplement*, p. 1; *Boomerang*, 15 December 1888, p. 13.

20. Ellis Rowan's award of a "first order of merit" is listed first of the Queensland awards in the official report on the Court in *Queensland Votes and Proceedings of the Legislative Assembly*, Second Session of the Tenth Parliament, 1889, vol. 2, p. 511.

1891

21. *Flower-Hunter*, p. 4.

22. R. G. Casey Senior was the father of Lord Casey, Governor-General of Australia from 1965 to 1969. See *Australian Dictionary of Biography*, vol. 3, 1851-1890, pp. 366-67; and Maie Casey, 'Mrs R. G. Casey, Senior' in *Rare Encounters*, Nomad Press, Melbourne, 1980, pp. 12-33.

23. 'Brisbane Society', *Queenslander*, 4 July 1891, p. 5.

24. 'Social Gossip', *Queenslander*, 25 July 1891, p. 149.

25. *Flower-Hunter*, p. 6.

26. See obituary, *Brisbane Courier*, 20 April 1925, p. 4. According to Henry Tardent (*Mrs Ellis Rowan and her Contributions to Australian Art and Science*, p. 8), Soutter and his wife Pattie became close friends of Ellis Rowan.

27. 'Personal', *Table Talk* (Melbourne), 14 August 1891, p. 2; *R. G. Casey Senior*,

Diaries, Entry for 1 August 1891, *Casey Papers*, MSS6150, series 4, NLA.

28. *Flower-Hunter*, pp. 8-10.
29. *Ellis Rowan Letters from Queensland*, 1891, letter, 14 August, p. 4.
30. Shipping departures, *North Queensland Herald* (Townsville), 19 August 1891, p. 29.
31. *Flower-Hunter*, p. 133.
32. See A. J. Harrison, 'William Saville-Kent, FLS, FZS, FRMS 1845-1908', *Papers and Proceedings of the Royal Society of Tasmania*, vol. 122, no. 2, 1988, pp. 165-78.
33. 'The Auxiliary Squadron. The Governor at Thursday Island', *Brisbane Courier*, 20 August 1891, p. 5.
34. See *Australian Dictionary of Biography*, vol. 4, 1851-1890, p. 471.
35. *Flower-Hunter*, p. 137.
36. *Ellis Rowan Cuttings Book*, pp. 9, 80. The journey to the Batavia River would have been in the company of the Jardines who also owned Bertiehaugh Station on the nearby Ducie River.
37. *Ellis Rowan Cuttings Book*, pp. 5, 12.
38. *Flower-Hunter*, p. 139.
39. *Ellis Rowan Letters from Queensland*, 1891, letter, 20 September, p. 25.
40. *Ellis Rowan Cuttings Book*, p. 3. Ellis described her visit to Jervis Island in *Flower-Hunter*, pp. 145-54.
41. Thursday Island Report for 24 September, *Brisbane Courier*, 25 September 1891, p. 5.
42. *Queensland Parliamentary Debates, of the Legislative Assembly*, vol. 64, 1891, 30 September, p. 1295.
43. See *Australian Dictionary of Biography*, vol. 4, 1851-1890, pp. 89-92.
44. *Flower-Hunter*, p. 135.
45. *Age* (Melbourne), 29 October 1891, p. 5.
46. *Ellis Rowan Cuttings Book*, p. 3. The exhibition catalogue has not survived.
47. Shipping arrivals, *Cooktown Independent*, 30 September 1891, p. 2.
48. *Ellis Rowan Letters from Queensland*, 1891, letter, October, p. 32.
49. Shipping arrivals, *Cairns Post*, 7 October 1891, p. 2.
50. *Flower-Hunter*, p. 39.
51. ibid, p. 40.
52. ibid, pp. 45-46. Ellis Rowan's journey can be dated to October 1891, since the first passenger train ran on the line in June that year.
53. 'Riding a CowCatcher. An Australian Artist's Adventures', *New Idea*, 6 January 1905, p. 641.
54. *Ellis Rowan Letters from Queensland*, 1891, letter, 20 October, p. 37.
55. *Flower-Hunter*, p. 50.
56. 'Queensland's Flora', *Daily Mail* (Brisbane), 3 August 1912, p. 2.
57. *Flower-Hunter*, p. 51.
58. *Ellis Rowan Letters from Queensland*, 1891, letter, 31 October, p. 47.
59. *Flower-Hunter*, p. 52. Ellis Rowan's hosts, John Stevenson and R. G. Casey Senior, were directors of the Muldiva Mining and Smelting Company. Another acquaintance, Charles Lumley Hill, was its local managing director. The company's shares were at a peak on the Brisbane Stock Exchange earlier in 1891, and it is likely that Ellis herself had shares. The mine did not fulfil its expectations and closed in 1893.
60. *Flower-Hunter*, p. 57.
61. ibid, p. 58.
62. ibid, pp. 61-65.
63. *Ellis Rowan Letters from Queensland*, 1891, letter, 5 November, pp. 54-61.
64. ibid, letter, 20 October, p. 45.
65. *Flower-Hunter*, pp. 68-95.
66. In particular, 'Herberton to Georgetown by Our Travelling Reporter', *Queenslander*, 5 September 1891, p. 451; reprinted *Brisbane Courier*, 8 September 1891, p. 2.
67. *Ellis Rowan Letters from Queensland*, 1891, letter, 5 November, p. 61.
68. Shipping arrivals, *Brisbane Courier*, 28 November 1891, p. 3.

1892

69. Shipping arrivals and departures, *Queenslander*, 4 June 1892, p. 1100, and *Northern Mining Register* (Charters Towers), 8 June 1892, p. 40. It was probably no coincidence that two acquaintances of Ellis Rowan, John Stevenson and Charles Lumley Hill, travelled on the same ship north from Brisbane. As directors of the Muldiva Mining and Smelting Company, they were on their way to the mine to report on recent breakdowns of the smelting machinery.
70. Richard Semon, *In the Australian Bush and On the Coast of the Coral Sea*, Macmillan and Co., London, 1899, p. 272.
71. *Flower-Hunter*, pp. 98-99.
72. See p. 35 of this publication.
73. Press clipping in *Anniechen MacArthur's Autograph Book*, held by Mr Peter Croker.
74. *Flower-Hunter*, pp. 102-06.
75. ibid, pp. 107-08.
76. ibid, p. 121.
77. ibid, p. 116.
78. *Argus*, 7 November 1892, p. 5.
79. Letter, Robert Bruce to Mrs Haddon, op. cit.
80. ibid.
81. ibid. Ellis Rowan sang the pawpaw's praises in her *Letters from Queensland*, 1891, third letter, p. 23.
82. 'Rare Flowers on Canvas', *Ellis Rowan Cuttings Book*, p. 5.
83. *Flower-Hunter*, p. 164.
84. ibid, p. 159.
85. Helen Topliss, *Tom Roberts 1856-1931. A Catalogue Raisonne*, Oxford University Press, Melbourne, 1985, Appendix I: 'Going North' by Tom Roberts, reprinted from *Argus*, November-December 1892.
86. *Flower-Hunter*, pp. 165-71.
87. Thursday Island Report for 3 September, *Brisbane Courier*, 5 September 1892, p. 5.
88. *Flower-Hunter*, p. 171.
89. Thursday Island Report for 15 September, *Queenslander*, 1 October 1892, p. 629.
90. Shipping arrivals, *Cooktown Courier*, 20 September 1892, p. 2.
91. 'Social Gossip', *Queenslander*, 22 October 1892, p. 773.
92. Shipping departures, *North Queensland Register* (Townsville), 12 October 1892, p. 18.
93. *Argus*, 7 November 1892, p. 5.
94. *Catalogue of Exhibits in the New South Wales Courts, World's Columbian Exposition, Chicago, 1893*, Government Printer, Sydney, 1893, item 1187, pp. 436-39. The paintings are listed by title, and include cat. nos 22, 46 and possibly cat. nos 41, 80.
95. 'Chicago Exhibition — New South Wales House', *Australasian Builder and Contractors' News*, 29 July 1893, p. 56.

Politicking in Queensland

1911

1. Letter, G. H. M. Addison to the Under Secretary of the Chief Secretary's Department, 14 December 1911, 1926/5028, PRE/A872, QSA.
2. 'Women's World', *Brisbane Courier*, 2 June 1911, p. 7; 'Personalities', *Queenslander*, 10 June 1911, p. 15.
3. 'Personal News', *Morning Bulletin* (Rockhampton), 17 June 1911, p. 8; 'Personal News', *Capricornian*, 24 June 1911, p. 33.
4. 'Personal News', *Morning Bulletin*, 22 June 1911, p. 5.
5. Shipping departures, *North Queensland Register*, 17 July 1911, p. 60.

6. Shipping arrivals, *North Queensland Register*, 9 October 1911, p. 53.

7. *Ellis Rowan Cuttings Book*, p. 54 (review of her Sydney exhibition of 1920).

8. 'Intrepid Painter of Flowers Describes Her Perilous Travels', *Ellis Rowan Cuttings Book*, p. 12.

9. Letter, Dr Carden Wallace to the author, 30 November 1989.

10. See C. T. White, 'F. M. Bailey: His Life and Work', *Proceedings of the Royal Society of Queensland*, vol. 61, no. 8, 1950, pp. 105-14.

11. F. Manson Bailey, 'Contributions to the Flora of Queensland', *Queensland Agricultural Journal*, vol. 28, January 1912, p. 74. The painting is reproduced as plate 15 of Bailey's *Comprehensive Catalogue of Queensland Plants*, (1913); and in Henry A. Tardent's *Mrs Ellis Rowan and her Contributions to Australian Art and Science*, 1927, p. 21.

12. Press clipping attached to letter, Ellis Rowan to the Queensland Premier Digby Denham, 6 December 1911, 1926/5028, PRE/A872, QSA.

13. See *Australian Dictionary of Biography*, vol. 7, 1891-1939, pp. 165-66.

14. *Ellis Rowan Cuttings Book*, p. 12.

15. See *Australian Dictionary of Biography*, vol. 8, 1891-1939, pp. 281-82.

16. Press clipping attached to letter, Ellis Rowan to the Premier, op. cit. The letter refers to *Banksia*, whereas the press clipping refers, inaccurately, to *Bauhinia*.

17. Letter, the Premier to the Under Secretary of the Department of Agriculture and Stock, 20 November 1911, 1926/5028, PRE/A872, QSA.

18. Letter, G. H. M. Addison to the Under Secretary of the Chief Secretary's Department, op. cit.

19. The colour plates include more close-up versions of eleven paintings in the Queensland Museum collection: cat. nos 13, 14, 36, 49, 52, 62, 105, 107, 110, 115, 122.

20. See S. T. Blake, 'Cyril Tenison White 1890-1950', *Proceedings of the Royal Society of Queensland*, vol. 62, no. 3, 1950, pp. 35-39.

21. F. Manson Bailey, 'Contributions to the Flora of Queensland', *Queensland Agricultural Journal*, vol. 28, March 1912, pp. 195-204.

22. For example, cat. nos 17, 20, 40, 78, 88, 106.

23. Letter, Ellis Rowan to the Premier, op. cit.

24. Letter, Ellis Rowan to the Queensland Premier Digby Denham, 23 December 1911, 1926/5028, PRE/A872, QSA.

1912

25. 'Social', *Brisbane Courier*, 11 July 1912, p. 9; *R. G. Casey Senior, Diaries*, Entries for July 1912, *Casey Papers.* The Caseys had left Queensland in 1893.

26. See entry for Mrs Lumley Hill's recently deceased husband in *Australian Dictionary of Biography*, vol. 4, 1851-1890, p. 398.

27. 'Social', *Brisbane Courier*, 20 July 1912, p. 7.

28. Letter, F. Manson Bailey to the Museum Scientific Assistant, H. A. Longman, 9 January 1913, *Queensland Museum Correspondence.*

29. A list of plant names which Bailey provided for Ellis Rowan in 1913 is also in C. T. White's writing (letter 236, *F. Manson Bailey Letter Book*, no. 10, 1913, Queensland Herbarium Archives).

30. Apparently Ellis had first offered her Queensland paintings to the Government in 1911 (Letter, G. H. M. Addison to the Under Secretary, op. cit). Her asking prices then and in 1912 are unknown.

31. 'Queensland's Flora', *Daily Mail*, 3 August 1912, p. 2.

32. Letter, Ellis Rowan to the Under Secretary of the Chief Secretary's Department, 2 August 1912, 1926/5028, PRE/A872, QSA.

33. 'Mrs Rowan's "At Home" ', *Ellis Rowan Cuttings Book*, p. 38. Significantly, the architect G. H. M. Addison, a member of the Art Gallery's Board of Advice, and the art master Godfrey Rivers, the Gallery's Honorary Curator, were the only notable local art figures known to attend.

34. Letter, Ellis Rowan to the Under Secretary, op. cit.

35. 'About the Town by Eyesopen', *Ellis Rowan Cuttings Book*, n.p.

36. *Brisbane Courier*, 16 August 1912, p. 11, and 21 August 1912, p. 11.

37. *Brisbane Courier*, 17 August 1912, p. 7, and 20 August 1912, p. 9.

38. Note by Mrs A. B. Wilson, 1 September 1939, in a copy of *Bill Baillie: His Life and Adventures*, presented to A. B. Wilson by Ellis Rowan on 9 September 1912, held by the Wilson family. Ellis also presented A. B. Wilson with two of her paintings in appreciation of his support.

39. See Donald Watson and Judith McKay, *A Directory of Queensland Architects to 1940*, Fryer Memorial Library Occasional Publication No. 5, University of Queensland Library, St Lucia, 1984, p. 207.

40. 'Queensland Wild Flowers', Letter to the Editor by A. B. Wilson, *Brisbane Courier*, 22 August 1912, p. 9; reprinted as 'Australian Wild Flowers', *Daily Mail*, 26 August 1912, p. 8.

41. 'Social Items', *Daily Mail*, 28 August 1912, p. 2.

42. 'Mrs Rowan's Paintings', *Brisbane Courier*, 31 August 1912, p. 11, and *Daily Mail*, 31 August 1912, p. 4.

43. 'The Rowan Pictures', Letter to the Editor by A. Norton, President of the Horticultural Society of Queensland, *Brisbane Courier*, 10 September 1912, p. 4.

44. 'The Rowan Paintings. An Appreciation' by Henry Tryon (identified only as H. T.), *Brisbane Courier*, 7 September 1912, p. 13. The entomologist Henry Tryon, possibly a companion on Ellis Rowan's 1911 expedition to the Macpherson Range, was also instigator of the Queensland Naturalists' Club's resolution of support passed at their meeting of 29 August.

45. 'Australian Flowers', Letter to the Editor by Sydney B. J. Skertchly, *Brisbane Courier*, 5 September 1912, p. 5.

46. 'Social Items', *Daily Mail*, 28 September 1912, p. 8.

47. Letter, the Under Secretary of the Chief Secretary's Department to Ellis Rowan, 14 September 1912, 1926/5028, PRE/A872, QSA.

48. 'Mrs Rowan's Pictures Purchased by the Government', *Brisbane Courier*, 16 September 1912, p. 6; 'Paintings Purchased', *Daily Mail*, 16 September 1912, p. 4.

49. *Ellis Rowan Cuttings Book*, p. 38.

50. 'The Rowan Pictures', Letters to the Editor by Dr H. M. Lightoller, *Brisbane Courier*, 25 September 1912, Courier Home Circle, p. 8, and 13 November 1912, Courier Home Circle, p. 8.

51. 'The Rowan Pictures', Letter to the Editor by Sarah Hynes, *Brisbane Courier*, 6 November 1912, Courier Home Circle, p. 6.

52. 'The Rowan Pictures', Letter to the Editor by Sarah Hynes, *Brisbane Courier*, 18 December 1912, Courier Home Circle, p. 7.

53. Letter, the Premier Digby Denham to A. B. Wilson, 20 September 1912, 1926/5028, PRE/A872, QSA.

54. 'The Rowan Pictures', Letter to the Editor by A. E. C., *Brisbane Courier*, 2 October 1912, Courier Home Circle, p. 8.

55. 'The Rowan Pictures', Letter to the Editor by G. H. B., *Brisbane Courier*, 17 September 1912, p. 8.

56. See *Australian Dictionary of Biography*, vol. 11, 1891-1939, pp. 397-98.

57. Letter, Godfrey Rivers to the Premier, 26 December 1911, 1926/5028, PRE/A872, QSA.

58. Shipping departures, *Brisbane Courier*, 28 September 1912, p. 6.

59. Letter, Ellis Rowan to Jane King, 19 September 1912, *Copies of Letters to Miss Jane King*, document 549, Mitchell Library.

60. Queensland Herbarium specimen nos AQ214897 and AQ341844, the latter described in the Addenda of Bailey's *Comprehensive Catalogue of Queensland Plants*, (1913), p. 836; and yet another specimen noted in the Addenda, p. 837.

61. 'Townsville Gossip', *North Queensland Register*, 9 December 1912, p. 17.

62. 'Cardwell Notes', *North Queensland Register*, 16 December 1912, p. 43.

63. Interview, Mr Peter Croker with the author, 14 June 1989.

64. 'Cardwell Notes', op. cit.

65. 'Townsville Gossip', op. cit.

66. Shipping arrivals, *Brisbane Courier*, 6 December 1912, p. 4.

67. 'Social Items', *Daily Mail*, 11 December 1912, p. 8.

68. 'Social Items', *Daily Mail*, 20 December 1912, p. 2.

69. Letter, Ellis Rowan to Jane King, n.d. — probably December 1912, *Copies of Letters to Miss Jane King.*

1913

69. 'Social', *Brisbane Courier*, 24 June 1913, p. 9.

70. *Queensland Figaro*, 10 July 1913, p. 10.

71. Letter, Ailsa to Jane King, 6 August 1963, *Copies of Letters to Miss Jane King.*

72. Shipping departures, *Brisbane Courier*, 12 July 1913, p. 6.

73. 'Out of It', *Queensland Art Society Cuttings Book*, 1909-1938, p. 111, Acc. 881004, series 16, item 1, Fryer Memorial Library, University of Queensland.

74. Letter, Mrs Doris Lakeland to the author, 24 June 1989.

75. List of painting titles for Ellis Rowan, letter 236, n.d., *F. Manson Bailey Letter Book*, no. 10, 1913, Queensland Herbarium Archives.

76. Letter, F. Manson Bailey to Ellis Rowan, letter 7, 6 October 1913, *F. Manson Bailey Letter Book*, no. 11, 1913.

77. Letter, Ellis Rowan to F. Manson Bailey, letter 8, 28 October 1913, *F. Manson Bailey Letter Book*, no. 11, 1913.

78. Shipping arrivals, *Brisbane Courier*, 16 September 1913, p. 6.

79. Letter, Ellis Rowan to F. Manson Bailey, op. cit.

Queensland Museum Collection

1. Letter, the Museum Director Dr R. Hamlyn Harris to the Under Secretary of the Chief Secretary's Department, 19 September 1912, *Queensland Museum Correspondence.*

2. Letter, the Under Secretary to Ellis Rowan, op. cit.

3. Ellis Rowan's exhibition at Angus and Robertson's Gallery, Sydney, in 1910 was criticised for inaccurate botanical titles *(Ellis Rowan Cuttings Book,* p. 30).

4. 'Mrs Rowan's Pictures', Letter to the Editor by M. L. P., *Brisbane Courier*, 11 September 1912, Courier Home Circle, p. 7.

5. See Jules L. Tardent, *Tardent Family History and Genealogy*, Tardent Family History Trust, Queensland, (1982).

6. The federal collection remained in storage at the Commonwealth Treasury until 1934 when it was transferred to the Parliamentary Library and thence to the NLA.

7. Henry A. Tardent, *Mrs Ellis Rowan and her Contributions to Australian Art and Science*, Watson and Ferguson, Brisbane, 1927, p. 32.

8. Interview, Miss Enid Tardent with the author, 2 June 1989.

9. Letter, the Museum Director Dr R. Hamlyn Harris to the Under Secretary of the Chief Secretary's Department, 31 October 1912, *Queensland Museum Correspondence.*

Techniques

10. For example, *Flower-Hunter*, pp. 10, 122.

11. ibid, pp. 6, 98.

12. Interview, Mr Peter Croker with the author, 14 June 1989.

13. 'Riding a CowCatcher. An Australian Artist's Adventures', op. cit, p. 641.

14. 'Rare Flowers on Canvas', *Ellis Rowan Cuttings Book*, p. 5.

15. Maie Casey, *An Australian Story*, p. 106.

16. Letter, Ellis Rowan to the Premier, 6 December 1911.

17. Wilfrid Blunt, *The Art of Botanical Illustration*, Collins, London, 1955, pp. 203-08.

18. Many of the duplicates are on green paper when the originals are on grey paper.

19. Margaret Hazzard, *Flower Paintings of Ellis Rowan*, NLA, Canberra, 1982, p. 14.

20. Ellis Rowan undertook a painting expedition to the northwest of Western Australia with Margaret Forrest in 1889.

21. The best known was an older contemporary, Louisa Ann Meredith (1812-1895) of Tasmania, whose paintings were more restrained and sentimental.

22. *Boomerang*, 15 December 1888, p. 13.

Plants

23. Letters, F. Manson Bailey to Ellis Rowan, letter 8, 7 October 1913, and Ellis Rowan to Bailey, 28 October 1913, *F. Manson Bailey Letter Book*, no. 11, 1913.

24. Letter, Ellis Rowan to the Premier, op. cit.

25. *Flower-Hunter*, p. 138.

26. ibid, p. 141.

27. In the garden at Macknade Plantation, for example, Ellis recognised "many old Colombo and Indian friends" *(Flower-Hunter*, p. 24).

Animals

28. New Guinea bird paintings, numbering 79, are a major component of the Ellis Rowan Collection in the NLA.

29. Semon, p. 272.

30. Maie Casey, *An Australian Story*, p. 105.

31. *Flower-Hunter*, p. 28.

32. ibid, pp. 36, 38. The new moth which Ellis accurately describes is clearly *Phyllodes imperialis* Druce.

33. *Flower-Hunter*, p. 20.

34. ibid, p. 14.

35. ibid, p. 97.

36. 'A Flower Hunter. Mrs Ellis Rowan's Travels', *Argus*, 10 December 1904, p. 17.

37. Margaret Hazzard, *Australia's Brilliant Daughter Ellis Rowan*, p. 133.

38. For example, Ellis added insects to cat. nos 39, 63 and 96 in her duplicate paintings at the NLA (R2265, R2379 and R2109 respectively); and deleted insects from cat. nos 76 and 93 in other duplicates (R2273 and R2532 respectively).

39. *Flower-Hunter*, p. 20.

SELECT BIBLIOGRAPHY

Books and Articles

A Fine Field for Botanizing. Mitchell and Dixson Galleries, State Library of New South Wales, Sydney, 1981 (exhibition catalogue).

A Vision of Eden: the Life and Work of Marianne North. Holt, Rinehart and Winston, New York, 1980, published in collaboration with the Royal Botanic Gardens, Kew.

AMBRUS, Caroline (comp.). *The Ladies' Picture Show: Sources on a Century of Australian Women Artists.* Hale and Iremonger, Sydney, 1984.

Australian Dictionary of Biography. Melbourne University Press, vols 3-5, 1851-1890; vols 7, 8, 11, 1891-1939.

BAILEY, F. Manson. *Comprehensive Catalogue of Queensland Plants.* A. J. Cumming Government Printer, Brisbane, (1913).

BLAKE, S. T. 'Cyril Tenison White 1890-1950'. *Proceedings of the Royal Society of Queensland,* vol. 62, no. 3, 1950, pp. 35-39.

BLUNT, Wilfrid. *The Art of Botanical Illustration.* Collins, London, 1955.

CASEY, Maie. *An Australian Story 1837-1907.* Michael Joseph, London, 1962.

CASEY, Maie. 'Mrs R. G. Casey, Senior' in *Rare Encounters.* Nomad Press, Melbourne, 1980, pp. 12-33.

COLLINS, Kate. *Ellis Rowan 1848-1922: A Biographical Sketch.* Oz Publishing, Brisbane, 1988.

GOODING, Janda. *Margaret Forrest: Wildflowers of Western Australia.* Art Gallery of Western Australia, Perth, 1984 (exhibition catalogue).

HARRISON, A. J. 'William Saville-Kent, FLS, FZS, FRMS 1845-1908'. *Papers and Proceedings of the Royal Society of Tasmania,* vol. 122, no. 2, 1988, pp. 165-78.

HAZZARD, Margaret. *Australia's Brilliant Daughter Ellis Rowan: Artist, Naturalist, Explorer 1848-1922.* Greenhouse Publications, Melbourne, 1984.

HAZZARD, Margaret

HEWSON, Helen. *Flower Paintings of Ellis Rowan from the Collection of the National Library of Australia.* National Library of Australia, Canberra, 1982.

KERR, Joan. Ellis Rowan entry in *200 Australian Women: A Redress Anthology,* ed. Heather Radi. Women's Redress Press, Sydney, 1988, p. 53-54.

KERR, Ruth S. 'Muldiva'. *Cairns Historical Society Bulletin,* 269, May 1982, pp. 1-3.

LOUNSBERRY, Alice. *A Guide to the Wild Flowers.* Frederick A. Stokes and Company, New York, 1899.

LOUNSBERRY, Alice. *A Guide to the Trees.* Frederick A. Stokes and Company, New York, 1900.

LOUNSBERRY, Alice. *Southern Wild Flowers and Trees.* Frederick A. Stokes and Company, New York, 1901.

MONTEITH, G. B. 'History of Biological Collecting at Cape York, Queensland 1770-1970'. *Queensland Naturalist,* 28, 1-4, 1987, pp. 42-51.

MOORE, William. *The Story of Australian Art.* Angus and Robertson, Sydney, 1934.

MOYAL, Ann. 'Collectors and Illustrators: Women Botanists of the Nineteenth Century' in *People and Plants in Australia,* ed. D. J. and S. G. M. Carr. Academic Press, Sydney, 1981, pp. 333-56.

NORTH, Marianne. *Some Further Recollections of a Happy Life, Selected from the Journals of Marianne North, Chiefly between the Years 1859 and 1869,* edited by her sister Mrs John Addington Symonds. Macmillan and Co., London, 1893.

PHIPPS, Jennifer. *Artists' Gardens: Flowers and Gardens in Australian Art 1780s-1980s.* Bay Books, Sydney, (1986).

Press Opinions of Mrs F. C. Rowan's Watercolour Flower Drawings. Stenlake and Simpson, London, (1905).

ROWAN, Ellis. 'Pen and Pencil Notes of a Trip through New Zealand' series in *Australian Town and Country Journal,* 23 December 1893 - 11 August 1894.

ROWAN, Ellis. *A Flower-Hunter in Queensland and New Zealand.* Angus and Robertson, Sydney, 1898.

ROWAN, Ellis. 'An Australian Artist's Adventures' series in *New Idea,* 6 January - 5 August 1905.

ROWAN, Ellis. *Bill Baillie: His Life and Adventures.* Whitcombe and Tombs, Melbourne, 1908.

ROWAN, Ellis. 'Making of an Art Collection'. *Oarswoman,* Brisbane, December 1912.

SAMUEL, H. J. *Wild Flower Hunter: The Story of Ellis Rowan.* Constable and Co., London, 1961.

SEMON, Richard. *In the Australian Bush and on the Coast of the Coral Sea.* Macmillan and Co., London, 1899.

STANLEY, T. D. and E. M. ROSS. *Flora of South Eastern Queensland.* Department of Primary Industries, Brisbane, vol. 1, 1983; vol. 2, 1986.

TARDENT, Henry A. *Mrs Ellis Rowan and her Contributions to Australian Art and Science.* Watson and Ferguson, Brisbane, 1927.

TARDENT, Jules L. *Tardent Family History and Genealogy.* Tardent Family History Trust, Queensland, (1982).

THORNTON, Robert John. *The Temple of Flora.* New York Graphic Society, Boston, (c.1981).

TOPLISS, Helen. *Tom Roberts 1856-1931. A Catalogue Raisonne.* Oxford University Press, Melbourne, 1985.

WHITE, C. T. 'F. M. Bailey: His Life and Work'. *Proceedings of the Royal Society of Queensland,* vol. 61, no. 8, 1950, pp. 84-86.

WILLIAMS, K. A. W. *Native Plants of Queensland.* Keith A. W. Williams, North Ipswich, vol. 1, 1979; vol. 2, 1984; vol. 3, 1987.

Official Records

Catalogue of Exhibits in the Queensland Court, Centennial International Exhibition, Melbourne, 1888. Warwick and Sapsford, Brisbane, 1888.

Catalogue of Exhibits in the New South Wales Courts, World's Columbian Exposition, Chicago, 1893. Government Printer, Sydney, 1893.

F. Manson Bailey Letter Books, nos 10 and 11, 1913, Queensland Herbarium Archives.

C. T. White Cuttings Book, Queensland Herbarium Archives.

Colonial Secretary's Office Correspondence, COL/A514, letter 1887/6819, Queensland State Archives.

Agent-General, London Dispatches, COL/128, letter 1895/11895, Queensland State Archives.

Premier's Department Correspondence, PRE/A872, file 1926/5028, Queensland State Archives.

Queensland Museum Correspondence, 1912-1913, 1924, Queensland Museum Library.

Queensland Museum Cuttings Book, 1914-1924, Queensland Museum Library.

Queensland Parliamentary Debates, of the Legislative Assembly, vol. 64, 1891, 30 September.

Report of the Queensland Commissioners of the Centennial International Exhibition, Melbourne, 1888. bound in *Queensland Votes and Proceedings of the Legislative Assembly,* Second Session of the Tenth Parliament, 1889, vol. 2.

Report of the Executive Commissioner for New South Wales to the World's Columbian Exposition, Chicago, 1893, Sydney, 1894, bound in *New South Wales Votes and Proceedings of the Legislative Assembly,* Session 1894-95, vol. 6.

Report by a Business Committee Appointed by the Commonwealth Government for the Purpose of Inquiring into the Value of the Late Mrs Ellis Rowan's Pictures, 27 January 1923, Parliament of the Commonwealth of Australia, 1923, Second Session.

Private Papers

Album of Photographs of Pictures by Ellis Rowan of the Pioneer, Herbert, and Johnstone Rivers and Thursday Island etc., presented to the Earl of Jersey, Mitchell Library.

Arthur Neame Diary, 1870-97, MSS 1122, Mitchell Library.

Casey Papers, Diaries of R. G. Casey Senior, 1891, 1892 and 1912, MSS 6150, series 4, National Library of Australia.

Copies of Letters to Miss Jane King from Ellis Rowan, document 549, Mitchell Library.

Ellis Rowan Cuttings Book and other papers, (MSS 2203, 2206, 2479, 2495, 422 and 2779), National Library of Australia.

Ellis Rowan Letters from Queensland, 1891, bound letter book held by the artist's family.

Haddon Collection, packet 1006, Cambridge University Library.

Jardine Cape York Collection, Royal Historical Society of Queensland Library.

Queensland Art Society Cuttings Book, 1909-38, Acc. 881004, series 16, item 1, Fryer Memorial Library.

Exhibition Catalogues

Catalogues of Ellis Rowan's exhibitions held by: the Pictorial Section of the National Library of Australia; the Art, Music and Performing Arts Library of the State Library of Victoria; the Mitchell Library; and the artist's family.

Newspapers and Journals

Newspapers and journals were searched for specific dates and localities. Information was retrieved from:

Age (Melbourne)
Argus (Melbourne)
Australasian (Melbourne)
Australasian Builder and Contractors' News
Australian Town and Country Journal

Boomerang (Brisbane)
Brisbane Courier
Bulletin

Cairns Argus
Cairns Post
Capricornian (Rockhampton)
Cooktown Courier
Cooktown Independent

Daily Mail (Brisbane)

Illustrated Sydney News

Lone Hand

Morning Bulletin (Rockhampton)

North Queensland Herald (Townsville)
North Queensland Register (Townsville)
Northern Mining Register (Charters Towers)

Queensland Agricultural Journal
Queensland Figaro
Queensland Naturalist
Queensland Post Office Directories
Queensland Society Magazine
Queenslander

Sydney Mail
Sydney Morning Herald

Table Talk (Melbourne)
Townsville Daily Bulletin
Townsville Herald.